Platform Interference
in Wireless Systems

Platform Interference in Wireless Systems

Models, Measurement, and Mitigation

Kevin Slattery & Harry Skinner

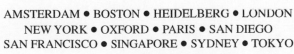

AMSTERDAM • BOSTON • HEIDELBERG • LONDON
NEW YORK • OXFORD • PARIS • SAN DIEGO
SAN FRANCISCO • SINGAPORE • SYDNEY • TOKYO

Newnes is an imprint of Elsevier

ELSEVIER

Newnes

Newnes is an imprint of Elsevier
30 Corporate Drive, Suite 400, Burlington, MA 01803, USA
Linacre House, Jordan Hill, Oxford OX2 8DP, UK

Library of Congress Cataloging-in-Publication Data
Application Submitted

British Library Cataloguing-in-Publication Data
A catalogue record for this book is available from the British Library.

ISBN: 978-0-7506-8757-7

For information on all Newnes publications
visit our web site at *www.books.elsevier.com*

Transferred to Digital Printing, 2011

Printed and bound in the United Kingdom

Contents

Preface ix

Chapter 1: Introduction 1
  1.1 Setting The Scene 1
  1.2 Wireless vs. EMI 2
  1.3 Addressing the Fundamentals 4
      1.3.1 Platform Interference: Describing the Problem 6
  1.4 Platform RF Snapshots 26
  1.5 Platform RF Sources: Summing It All Up 34

Chapter 2: The Structure of Signals 37
  2.1 Time has Little To Do with Infinity and Jelly Doughnuts 37
  2.2 Platform Signals: Preliminaries in Spectral Analysis 38
  2.3 Harmonic Variation with Time 40
  2.4 Variation with Duty Cycle 43
  2.5 Nonrepetitive Signals 55
  2.6 Modifying a Clock 57
  2.7 Random Data Sequences as Signals 59
  2.8 Spectra of Display Symbols 66
  2.9 Summary 75
    References 76

Chapter 3: Analysis of Symbols 77
  3.1 When People Finally Come to Understand your Vision... It's Time for
      a New Vision 77
  3.2 Analysis of Display Symbols Part II 92
      3.2.1 Justification for the Differential Operator 102
  3.3 Wireless Performance in the Presence of Radiated Emissions from
      Digital Display Symbols 104
  3.4 Developing the Analysis of Differential Symbols 107

3.5   The Impact of Pulse Width and Edge Rate Jitter on the Expected EMI/RFI ... 113
3.6   Summary 119
References *119*

Chapter 4: Measurement Methods 121

4.1   One Measurement is Worth a Thousand Opinions 121
4.2   Near-field Scans of Clock ICs 128
4.3   Measurements Beyond the Near Field: Transition Region 133
4.4   The Near-Field, Far-Field Transition 133
4.5   Far-field Measurements 141
4.6   Other Measurement Methods 142
4.7   Summary 145
References *145*

Chapter 5: Electromagnetics 147

5.1   He Must Go by Another Way Who Would Escape This Wilderness 147
5.2   The Time-Varying Maxwell's Equations 154
5.3   Electric and Magnetic potentials 156
5.4   Radiation Mechanisms 158
References *163*

Chapter 6: Analytical Models 165

6.1   Honest Analysis Gets in the Way of Results Desired Emotionally 165
6.2   The Electric Dipole 165
6.3   Field Impedance of the Elementary Electric Dipole 172
6.4   The Extended Dipole Radiator 174
6.5   The Magnetic Dipole 180
6.6   Developing Simple Analytical Structures for Real Devices 185
6.7   Developing Channel Models 196
6.8   Summary 210
References *210*

Chapter 7: Connectors, Cables, and Power Planes 213

7.1   Chance and Caprice Rule the World 213
7.2   Power Distribution Radiated Emissions 215
7.3   Investigating the Radiated Emissions Potentials of Power Distributions
      in Packages and Silicon 229
7.4   Further Investigations with Various Power Topologies 237
      7.4.1 Substrate Noise 245
7.5   Summary 251
References *251*

Chapter 8: Passive Mitigation Techniques.......................**253**
8.1 Introduction 253
8.2 Passive Mitigation 254
  8.2.1 Shielding.....................................254
  8.2.2 Absorbers 276
  8.2.3 Layout 281
References...289

Chapter 9: Active Mitigation **291**
9.1 Frequency Planning 291
9.2 Frequency Content 297
  9.2.1 Spread Spectrum Clocking 301
9.3 Radio Improvements 305
  9.3.1 Noise Estimation 306
9.4 Closing Remarks 308
References...309

Appendix **311**

Index **329**

Preface

This book has been written with the practicing EMI engineer in mind. To that end, we have eschewed much of the usual mathematical development, providing only the aspects of theory that we think will help to advance the story and provide a means for readers to develop their own intuitions and perspectives. Where necessary, we have laid out the mathematics we used to develop our ideas and approaches. We have included in the appendices all code that we used toward model building and analysis. We have primarily relied upon Mathematica (Wolfram Research) and SystemView (Agilent). We assume that some readers will use the code to write their own routines in other systems.

For a long time, this work has been in piecemeal gestation for both authors. The entire work came together rather quickly in the end, and we wish to make it clear that any errors in text are purely our own.

CHAPTER 1

Introduction

1.1 Setting The Scene

With the advent of mobile computing, wireless communication has become an integral part of the computer platform. Who would now consider buying a laptop without wireless? At the same time, the once simple communications devices such as cell phones now have functions that require subsystems ordinarily associated with computer devices. So what's the big deal? The problem is that these devices were never intended to coexist. Communications devices were not designed with high-speed digital logic in mind. High-speed digital logic never included communications as a design vector. The end result is that these devices don't work well together, so much shoehorning is currently undertaken to make them cohabit in the same device. This shoehorning generally incurs costs in terms of product delays and additional mitigation solutions. It is a sobering thought that 3 dB of noise can reduce the performance of your communications system by 50%. It is even more sobering that 20 or even 30 dB of noise is common on some devices. This book has two main intentions:

1. To provide an education about the principles of radio frequency interference (RFI).

2. To provide a reference source for identifying noise-related issues and mitigating them in your current or future system design explanation.

When considering what we should name this book, many titles came to mind. We wanted to capture the uniqueness of the topic while communicating the essence of the book. Interference as a topic is overly broad. In the EMI/EMC world, interference is a system-to-system or environment-to-system phenomenon. Similarly, in the wireless world, interference can be one wireless system to another or a consequence of environment. Electromagnetic interference, or EMI, is measured in field strength (dBμV/m, or decibel ratio referenced to 1 microvolt per meter). In wireless, receive signal strength is a measure of power and is calculated using dBm (or decibel ratio of watts to 1 milliwatt). It is interesting to note that these two seemingly distinct and separate disciplines are so similar yet so different. This book is an attempt to bring these worlds together. You may ask why this is necessary and why now. The answer is quite simple.

In the mid 1990s, computer systems (classified as *unintentional radiators* by the Federal Communications Commission [FCC]) operated in the hundreds of MHz, and wireless functionality as part of a computer wasn't even a glimmer in someone's eye. Laptops were a fairly new device—an evolutionary development of the "luggables" of the early 1990s. Wireless communication (as a mobile feature/device) was essentially limited to cell phones, which were similar in many ways to these luggables. However, this has changed in the last few years. With the launch of Intel's Centrino® brand, wireless functionality became a fundamental part of mobile computing—a must-have. Today, many devices have wireless functionality integrated with computer functions. They range from GPS receivers on our vehicles that we can program for directions to handheld devices that combine a phone, a music player, and an instant and e-mail messaging system. It is clear that this trend will continue and is likely to accelerate. Designing these devices has become increasingly more complex. Although many industry experts talk about the convergence of computers and communications, it is clear that these are still two distinct worlds, and that the devices requiring both functions continue to challenge designers around the world. Why is this so difficult? Today's multiusage mobile devices incorporate one or more radios operating from 800 MHz to the single-digit GHz. They also incorporate digital circuitry with processing, memory, and storage operating from the high hundreds of MHz to several GHz. This overlap in frequency is the crux of the issue. Today's radios were not designed with this in mind. RF engineers typically design and simulate the operation of their radios in white or Gaussian noise environments. Today's digital systems were conceived without any comprehension of wireless communications, and they tend to generate noise that is inherently non-Gaussian. When these two components meet, we have an undefined environment that typically leads to less than stellar operation and below par performance. In some cases, wireless will simply cease to function. What we truly have is a collision between communications and computing. Special attention to this issue will ensure that devices requiring computer and wireless functionality make it to market in the desired timeframe at the required price points.

1.2 Wireless vs. EMI

One may ask why EMI regulations do not protect against this type of issue. To answer this question, we must first look at the intent as well as the context of the regulations. When such regulations first appeared, the FCC was concerned primarily with protecting terrestrial broadcasts such as TV and radio. Their major concern was that other devices in close proximity would interfere and disturb the reception of such services. The EMI limits introduced were therefore based on a system-to-system interference model. Hence, they were measured at a distance of 3 m, which, among other reasons, was to equate to device (aggressor-to-victim) separation. The limits were indeed intended to provide some level of protection from devices separated by approximately 10 feet, which at a stretch were located in the same room, but likely in an adjacent room or, in fact, next door. So, although perhaps a good starting point, it is clear that these regulations are wholly insufficient in themselves and in some ways completely inapplicable to protection of wireless devices collocated in the same system. In fact, the FCC doesn't care if you interfere with yourself! To illustrate this point, we'll look at the existing EMI limits (FCC Part 15 ITE) at 2.5 GHz and compare this to 802.11 b/g wireless sensitivity requirements. For the purpose of this illustration, we will assume that the radio receiver is 3 cm away from the EMI source.

EMI Limits:

2.5 GHz @1 MHz BW = 54 dBμV/m (@3 m)

Translating for distance = 94 dBμV/m (@3 cm)   [20 log (300/3)]

Converting to dBm = −13 dBm   [dBm = dBμV − 107]

Wireless Sensitivity Requirements:

802.11 b/g (11 Mbps) = −86 dBm (@20 MHz)

Converting for BW = −73 dBm (@1 MHz) [10 log (20/1)]

As we can see, there is a 60 dB difference between the EMI limit and the required level to ensure wireless functionality as indicated in the 802.11 specification. In the EMI world, 60 dB is a big number. In the wireless world, it is huge, where 1 and 2 dB performance gains can be the subject and result of a life's work. In real-world terms, this means that the wireless voltage requirements are 1,000 times more stringent than current EMI regulations. It is also clear that, as stated earlier, meeting EMI requirements is a good starting point at best. This may be a relatively simple model, but it nevertheless enforces the message that EMI/EMC compliance guarantees nothing with respect to wireless functionality when the radio is integrated into the same device as the digital electronics.

To further hit the point home, Figure 1.1 shows a plot of noise measured using 802.11 embedded antennas on a production notebook running between 2.3 and 2.6 GHz.

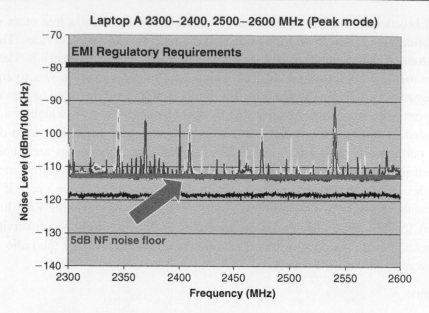

Figure 1.1: Noise measurements on a production notebook.

Two lines are plotted: the wireless sensitivity requirements (assuming a 5 dB noise figure for the radio and incorporating the gain for the antenna used) and the EMI limit for the device in question (FCC Class B). The system in question easily meets EMI requirements, but has considerable noise up to 20 dB above the level required by the integrated 802.11 radio. It is clear that this system may have difficulty operating to specified parameters, depending on what radio channel is used.

1.3 Addressing the Fundamentals

In discussing RFI and EMI in this book, our approach is to look for underlying physical phenomena in order to develop design techniques that can be applied consistently across multiple designs in a cost-effective manner. It is not our intention to help the reader identify point solutions that typically work for a single design instantiation and result in "new" approaches time and time again. In the same way, throwing parts at problems can work when the budget is unlimited, but doing this will cause problems when cost is a consideration. Besides, in most cases, this method treats the symptoms rather than the cause. It is the inherent intention of this book to address RFI fundamentals, solving the problem at the source where possible, or at least building the paddock as close to the barn as possible.

Correcting EMI and RFI problems *after* systems are designed and ready to go into production is usually expensive and can result in program delays that adversely affect the acceptance of a new product. It is preferable to follow good engineering practice during the design and development phases. The goal should be to produce designs capable of functioning without adverse effects in the predicted or specified electromagnetic environment. In addition, these devices must not interfere with other circuits. The electromagnetic theory, signal analysis, measurement techniques, compliance standards, and design guidelines described in this book should all aid in meeting the goal of intended wireless functionality and maximum system performance when applied early enough in the design cycle.

In a well-controlled RFI/EMC management scheme, the appropriate engineer reviews and approves all design drawings, takes part in design reviews, is a member of design change committees, and is notified of any proposed changes in a design. Keep in mind that any shortcut in EMC design or control can cause a lengthening in testing, fixing, and retesting time, as well as an additional cost.

The designer must realize that EMC is an iterative process composed of design, analysis, fabrication, and measurement. A first-pass solution is not always achieved. During project schedule development, the engineer should build into the process a sufficient amount of time to cover several iterations of EMC reduction.

In order to understand the best approach to an RFI/EMI solution, we must first define radiating structures and signal sources. Throughout the text, we will describe experiments and methods of measurement as they are applicable and in relation to the particular problems being addressed. The measurement techniques that we use will be discussed in detail. We will emphasize good measurement and experimental practices and not rely only on "clean" simulations. While measurement can be tricky, messy, and fraught with the possibility of error, it is still the very foundation of science.

In the first chapter, we will describe the problem of platform interference in wireless systems. We will show system spectrums of various types—the nature of the emissions of several dominant system components such as processors, chipsets, LAN, and so on. We will discuss the impact of even low-level emissions on wireless system performance. In Chapters 2 and 3, we will describe the analysis of signals in platforms such as high-speed data streams, system clocks, and display streams. In Chapters 2 and 3, we will cover measurement and modeling techniques. We will describe methods used to characterize the impact of different platform signal types on wireless performance. We will also discuss methods used to visualize the electric and magnetic fields in the near fields of radiators.

We will develop analytical approaches with the use of simulation tools, simple closed form analytical models, and the statistics of interferers. We will then show models of notebook display lids, desktop enclosures, packages, and heat sinks. In Chapter 4, we will outline and sketch approaches to mitigation of platform interference.

We will also discuss a number of commercially available analytical tools. We have leaned heavily on Mathematica from Wolfram Research, HFSS from Ansoft, and Flo-EMC from Flomerics. Each of these packages has strengths and complements the others.

1.3.1 Platform Interference: Describing the Problem

In order to address the challenges of RFI, one must take an approach to break the problem into manageable pieces. We will attempt to look at the problem from the perspective of both the digital electronics (the aggressor/source) as well as the radio (the victim). Figure 1.2 shows the problem in simple interference and electromagnetic compatibility terms.

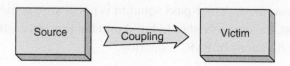

Figure 1.2: The aggressor/victim model.

The source above is the digital electronics and its subsystems. Examples are clocks, high-speed I/O (Input/Output) traces, cables, power delivery networks, LCD panels, memory subsystems, or processing components. The victim is the wireless radio itself and constitutes both the antenna subsystem and any associated RF components. The coupling depiction indicated above covers both radiated (through the air) and conducted (shared power, Vcc, and/or ground, Vss, connections). Typically, the coupling path represents a loss path that reduces the amount of energy the victim radio will see. Locating the antenna away from noisy electronics or shielding the noisy electronics increases this loss path, reducing the level of noise seen by the radio. The following example illustrates the power of this simple model in breaking down the problem.

Source X creates a noise equivalent to 10 μW (−20 dBm); 10% of this noise ends up being radiated into the surrounding environment (1 μW = −30 dBm). We will assume that there is some shielding between the noise source X and the radio antenna that gives us 40 dB attenuation. We can then calculate that the radio will see –70 dBm noise (–30 dBm—40 dB). This may seem okay, but given that the majority of radios have sensitivity in the –90 dBm range, this noise is already 20 dB above the sensitivity threshold of the radio, and the noise would reduce its performance by 75%. This may seem surprising. Even with 40 dB of isolation or attenuation between a 1 μW radiated noise source and a representative radio receiver, 75% of the wireless performance will be lost. It is even more surprising that only 10 pW of noise seen by the radio antenna can reduce the wireless performance by as much as 50% (range/throughput).

A very similar model, which looks at the problem from an RF perspective, is shown in Figure 1.3. The figure shows a rather simplified view of the radio world: the transmitter, the transmission channel, and the receiver. For now, we will simply look at the portion of the radio that resides on the platform and the part of the radio transmission channel associated with the platform. Platform noise is added into the channel at a number of possible locations and can be considered as the summation of individual sources resulting in a complex signal.

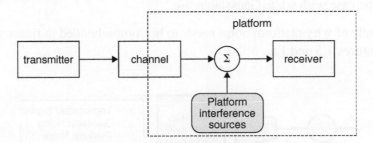

Figure 1.3: Signal channel and receiver.

A good representation of how platform noise adds to the transmitted radio signal such that the receiver sees a summation of the intended signal and the local generated platform noise. Unfortunately, this is not the model used by RF designers today; they use a very similar model, but they use a white Gaussian noise source instead of "Platform Interference Sources." The issue here is that platform-generated noise is typically non-Gaussian, and, as such, the present radio design does not comprehend the impact of such sources. Figure 1.4 illustrates this fact.

A theoretical representation of white Gaussian noise using normalized frequency is shown (*right*). The response is flat across the entire spectrum. Conversely, the figure on the left shows the noise from a 5 Gb/s Serial I/O bus, where the spectrum is far from flat with a

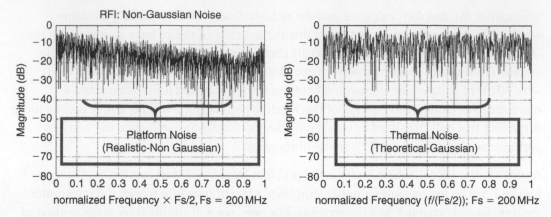

Figure 1.4: Measured PCIe Gen2 RFI emissions vs. Gaussian noise.

15 dB variation in peak amplitude. The net outcome is that the statistics related to the platform noise source now vary due to signal-to-interference (SIR) levels and bandwidth, which is not the case with white Gaussian noise.

Another example of why platform noise needs to be comprehended in future radio designs is shown in Figures 1.5 and 1.6.

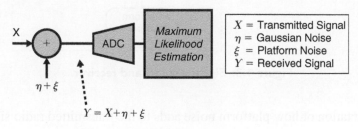

Figure 1.5: Radio platform channel model.

Figure 1.5 shows a realistic platform channel model incorporating both Gaussian and non-Gaussian (platform noise) components. Also shown is a representation of a maximum likelihood detector used to determine if a valid signal is being received by the radio. Figure 1.6 shows the resulting statistics seen at the analog-to-digital converter (ADC). It does not follow the expected normal distribution, therefore a maximum likelihood (ML) detector, which does not comprehend platform noise, will exhibit a higher probability of false alarm and/or missed detection.

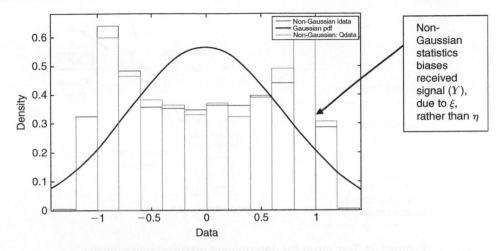

Figure 1.6: Received signal at ADC with and without platform noise.

1.3.1.1 Measuring Noise Impact

In the previous section, we covered basic approaches to breaking the problem into manageable pieces. We introduced a very simple noise budget model and identified issues with how current radios are designed today. To add a dose of reality to these topics, we will look at some data from production notebooks showing the existence of RFI issues today.

Figure 1.7 shows a setup used to identify the true impact on an embedded wireless device from platform-generated noise. Ordinarily it is not possible to turn off the digital circuitry and maintain wireless functionality, but the setup shown here allows it.

We need two identical systems for the test. The embedded antenna in the "interference" system is connected to the wireless card of the "client" system. An attenuator is placed at the output of the Access Point (AP) to reduce the transmitted and/or received signal to approximate for moving the client and AP further apart. In this way, we can test the comparable range and throughput performance. The wireless client is placed in a shielded room to ensure that the antenna in the interference system only sees noise from the interference system itself. Now we can turn the interference system (Laptop Model A #2) off and on as well as test different operating conditions to see if the wireless functionality is indeed impaired and under what conditions. Figure 1.8 shows a typical output from such a measurement. In this case, two tests were done using the 802.11b radio in Channel 2.

A baseline with the interference test system turned off (gray line) shows the performance of the radio without platform noise. The radio performance starts to decrease when the

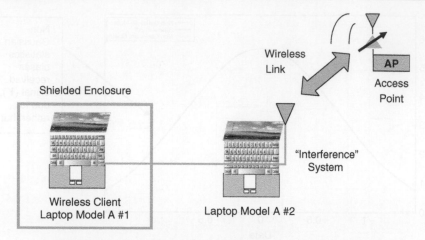

Figure 1.7: Test setup for determining platform noise impact.

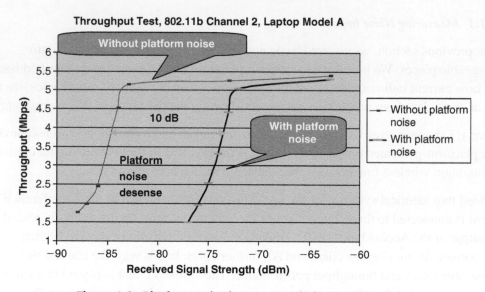

Figure 1.8: Platform noise impact on wireless performance.

received signal strength approaches –85 dBm, which is reasonably close to the specified performance of the embedded radio. The same test is repeated with the interference test system turned on (heavy curve-line). Platform noise from the interference system causes an approximate 10 dB impact or de-sense of the radio receiver.

Figure 1.9 shows an additional example of a different system with an 802.11g radio.

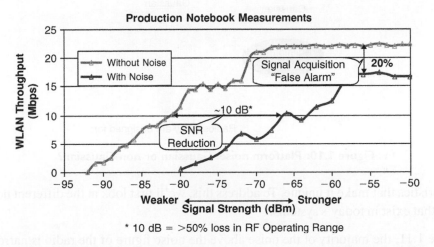

Figure 1.9: Production notebook measurements: impact of platform noise.

These measurements show two distinct impacts from platform noise. The signal-to-noise ratio (SNR) reduction is similar to that shown earlier in the chapter. This is the impact from the noise generated by the platform directly impacting the SNR of the radio receiver. The signal acquisition or "false alarm" impact is due to the misinterpretation of platform noise as a valid wireless signal (and the resultant overhead incurred in trying to process what the radio believes is a valid signal). In the meantime, a valid preamble signal comes along and the receiver misses it. In this example, a particular radio did not include platform noise and its specific characteristics as a design consideration.

1.3.1.2 *Platform Noise Characteristics*

We have asserted that platform-generated noise is non-Gaussian. We have also asserted that radios today typically only consider Gaussian noise in their design. Figure 1.10 shows a depiction of this.

As shown in Figure 1.10, platform-generated noise is a subset of the non-Gaussian category. Characterizing non-Gaussian noise is the ongoing subject of many theses, and both industry and academia will continue to pour considerable effort into its understanding. Platform noise may be non-Gaussian, but it does differ from what most theoreticians think of as non-Gaussian. Unlike most non-Gaussian noise, platform-generated noise has known

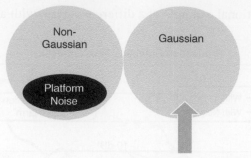

Figure 1.10: Platform noise: Gaussian or non-Gaussian?

characteristics that make it unique. To address this, we'll first look at the different noise sources that exist in today's systems.

In Figure 1.11, the majority of the noise above the noise figure of the radio is narrowband in nature. Although there is a broadband component, this noise is at or below the radio noise figure so will not affect the radio performance. In Figure 1.12, we see a much different spectrum.

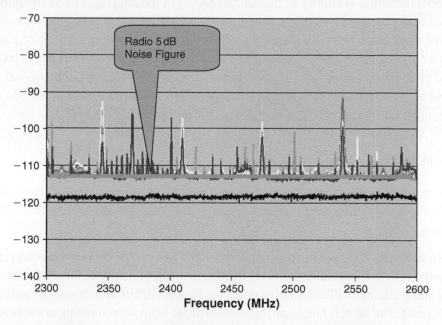

Figure 1.11: Narrowband noise in a production notebook.

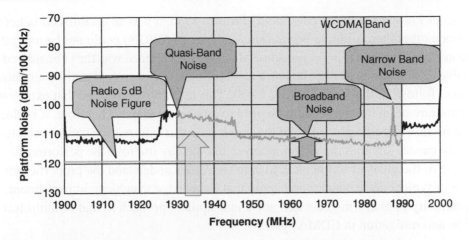

Figure 1.12: Wideband CDMA noise in a production notebook.

In Figure 1.12, we have zoomed in to a specific wideband Code Division Multiple Access (CDMA) radio frequency range. We can see three different noise signatures. As in Figure 1.11, the 5 dB NF line is identified. In this instance, there is indeed a broadband component that will affect radio performance. We also have one narrow band spike in the radio band. The arrow in the figure is highlighting a noise characteristic that is very unique to the digital systems of today. In this context, we will call it *quasi-band noise*. It is a direct result of the use of Spread Spectrum Clocking (SSC), also known as *dithered clocking*, in the system. It is, in fact, a narrow band clock noise signature that has a time-varying component. SSC is a well-known technique used to reduce EM emissions for passing regulatory EMC requirements. We will cover more about SSC later in this chapter and in subsequent chapters.

We therefore have three noise categories:

1. *Narrowband*: Noise with a spectrum of one or more sharp peaks that is narrow in width compared to the bandwidth of the victim radio receiver.

2. *Broadband*: Noise with a spectrum broad in width compared to the bandwidth of the victim radio receiver.

3. *Quasi-band*: Noise with a spectrum greater than, equal to, or less than the bandwidth of the victim radio receiver; the result of frequency modulation of a narrowband clock signal.

So how does knowledge of the noise help? Prior to 1986, it was an established belief that interference other than Additive White Gaussian Noise (AWGN) could not be mitigated. Detection of a radio signal in the presence of strong interference was thus considered impossible. This belief was dispelled as a result of intense research of multi-user detection during and following the 1990s (Verdu, 1998). This research established that exploitation of the structure of an interferer could be used to combat its effect. Therefore, it is clear that if you know specific properties of the noise, then that information can be used to essentially cancel it within the victim receiver and thereby improve the performance of the radio. One of the goals of this book is to help the reader understand the properties and structure of typical noise sources and apply that knowledge to achieve improvements in radio immunity to platform-generated noise in a similar approach to that of multiuser detection and mitigation in CDMA systems.

1.3.1.3 Spread Spectrum Clocking (Clock Dithering)

The following section will explain in detail the specific attributes of Spread Spectrum Clocks. As mentioned earlier, SSC is essentially a frequency modulation of a fixed frequency clock. This is illustrated in Figure 1.13.

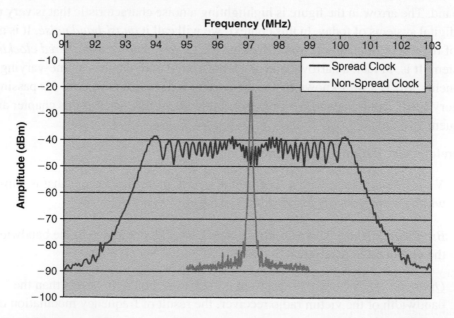

Figure 1.13: Spectral Fundamental Frequency comparison of SSC and non-SSC clocks.

The two plots shown in Figure 1.13 represent a non-spread 97.1 MHz clock and the same base clock with a center spread of approximately 8%. It is known as *center spread* as the modulation causes the clock to vary both above and below the nominal frequency. Another typical implementation results in what is known as *down spread*. Figure 1.14 shows similar plots for a down spread case.

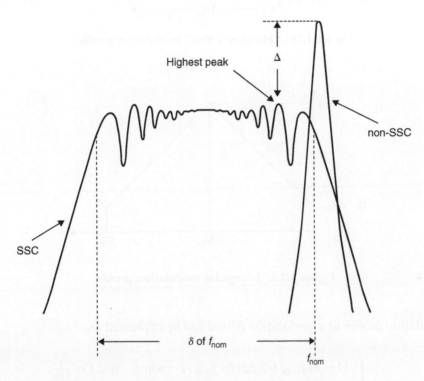

Figure 1.14: Spectral fundamental frequency comparison of SSC and non-SSC clocks.

In the down spread case, the spread modulation results in a clock that does not exceed the nominal frequency; it varies between the nominal frequency and a frequency that is lower than the nominal. In most digital systems, this lower frequency is approximately 0.5% lower than the nominal, and results in what is known as a $+0/-0.5\%$ spread clock. Also shown is the SSC EMI benefit Δ. There are two modulation profiles typically used in spread spectrum clocks: the "Hershey's Kiss" profile looks like its namesake, and a simple triangular profile. Both profiles are shown in Figures 1.15 and 1.16.

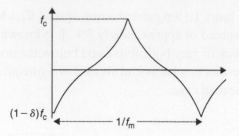

Figure 1.15: "Hershey's Kiss" modulation profile.

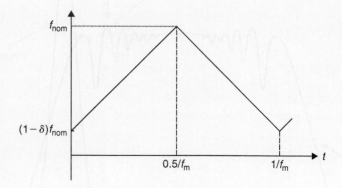

Figure 1.16: Triangular modulation profile.

The modulation profile in a modulation period can be expressed as:

$$f = \begin{cases} (1-\delta)\,f_{nom} + 2f_m \cdot \delta \cdot f_{nom} \cdot t & \text{when} \quad 0 < t < \frac{1}{2f_m}; \\ (1+\delta)\,f_{nom} - 2f_m \cdot \delta \cdot f_{nom} \cdot t & \text{when} \quad \frac{1}{2f_m} < t < \frac{1}{f_m}, \end{cases}$$

where

f_{nom} is the nominal frequency in the non-SSC mode,
f_m is the modulation frequency,
δ is the modulation amount, and
t is time.

The modulation frequency of SSC is typically in the range of 30 to 33 kHz to avoid audio band demodulation and to minimize system timing skew.

To understand how SSC may impact wireless communications, a fundamental understanding of both the time and frequency domain is required. Due to the response of the measurement equipment, illustrated in Figures 1.13 and 1.14, it would seem that the SSC signal continuously occupies a significantly larger spectrum than a non-SSC signal. This supposition is however incorrect. To explain this, we will consider the time domain. If we were to take a snapshot of the signal at any given instant in time, it would look very much like the non-SSC signal. At an incremental snapshot in time later, the SSC signal will occupy a similar bandwidth but at a different center frequency. By dithering the clock, the frequency of the fundamental and its harmonics will vary with time over a total spectrum consistent with the spread percentage. The amount of time spent at any one frequency will depend on the SSC modulation implementation in terms of signature (linear, or "Hershey's Kiss") and frequency (kHz).

1.3.1.4 Establishing Radio Frequency Interference/Platform Noise Risk

We will now return to the concept of a noise budget and establish a design process to establish RFI risks in your design. We'll start by determining the absolute radio performance (the best it could possibly be) and then show how radio sensitivity degrades as noise is added to the input. In the real world, noise cannot be avoided; it's everywhere, always. We can, however, recognize specific types and sources of noise, categorize them, and develop mitigation strategies to reduce the impact of these platform noise sources in selected radio bands. We cannot hope to reduce all sources of platform noise, but we don't have to. We only need to reduce the noise in the radio channels that cohabit the platform.

So, how do we approach quantifying these effects? Figure 1.17 is a schematic of additive noise. The thermal noise floor is the absolute lowest noise level that you can physically have and is measured with a bandwidth of 1 Hz. The noise floor of a receiver degrades from there. As receiver bandwidth increases, more noise is integrated. After bandwidth, the addition of the noise floor of the receiver comes. We then add in typical platform noise floor levels. So, when all is said and done, we have de-sensed the receiver noise floor from $-174\,\text{dBm/Hz}$ to $-77\,\text{dBm/MHz}$. Quite a change. However, the important point to note is that the receiver sensitivity without platform noise is $-91\,\text{dBm}$. In this example, the platform de-senses the receiver by 14 dB.

The table in Figure 1.18 shows allowable interference levels for a selection of well-known radios. It is calculated using the following equation:

Allowable Interference Level = Thermal Noise Density + Receiver Bandwidth
$$+ \text{Receiver Noise Figure} + \text{Desensitization Offset}$$

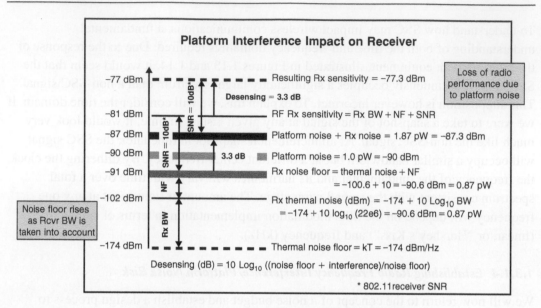

Figure 1.17: Schematic of receiver additive noise.

	Wi-Fi		
Noise Calculation	**802.11b**	**802.11g**	**802.11a (5 GHz)**
Thermal noise	−173.8 dBm	−173.8 dBm	−173.8 dBm
Channel B/W	73.4 dB	73.0 dB	72.2 dB
Thermal noise @ B/W	−100.4 dBm	−100.8 dBm	−101.6 dBm
Noise Figure	5.0 dB	5.0 dB	5.0 dB
Margin	0.0 dB	0.0 dB	0.0 dB
De-sensitization	10.0 dB	10.0 dB	10.0 dB
Allowable Interference Level (dBm)	**−105.4 dBm**	**−105.8 dBm**	**−106.6 dBm**

	WPAN		Satellite
Noise Calculation	**Bluetooth (2.4 GHz)**	**UWB (3–5 GHz)**	**GPS (1.55 GHz)**
Thermal noise	−173.8 dBm	−173.8 dBm	−173.8 dBm
Channel B/W	60.0 dB	87.0 dB	63.0 dB
Thermal noise @ B/W	−113.8 dBm	−86.8 dBm	−110.8 dBm
Noise Figure	15.0 dB	6.0 dB	3.0 dB
Margin	0.0 dB	0.0 dB	0.0 dB
De-sensitization	10.0 dB	10.0 dB	10.0 dB
Allowable Interference Level (dBm)	**−108.8 dBm**	**−90.8 dBm**	**−117.8 dBm**

	GPRS (2.5G)	EDGE (3G)	UMTS (3G, WCDMA)	CDMA (2G)
Noise Calculation	**GSM (850/900 MHz)**	**DCS/PCS (1.8/1.9 GHz)**	**DCS/PCS (1.8/1.9 GHz)**	**CDMA (925 MHz)**
Thermal noise	−173.8 dBm	−173.8 dBm	−173.8 dBm	−173.8 dBm
Channel B/W	53.0 dB	53.0 dB	67.0 dB	61.0 dB
Thermal noise @ B/W	−120.8 dBm	−120.8 dBm	−106.8 dBm	−112.8 dBm
Noise Figure	6.5 dB	7.0 dB	8.5 dB	7.0 dB
Margin	0.0 dB	0.0 dB	0.0 dB	0.0 dB
De-sensitization	10.0 dB	10.0 dB	10.0 dB	10.0 dB
Allowable Interference Level (dBm)	**−124.3 dBm**	**−123.8 dBm**	**−108.3 dBm**	**−115.8 dBm**

Figure 1.18: Allowable interference levels in wireless bands.

Thermal noise density is simply KT noise, or –173.8 dBm/Hz (defined at room temperature); Receiver bandwidth is defined by the specific radio standard. (For example, Bluetooth radio receiver's bandwidth is 1 MHz and 802.11b is 22 MHz.) To ease the calculation, the bandwidth is converted to dB using 10 log (BW); receiver noise figure is radio implementation-dependent and will vary from one radio to another as well as between manufacturers. Typical noise figures are used in the calculations above; de-sensitization offset refers to how much lower the interference signal needs to be below the receiver's own noise floor in order to achieve a specific level of "aggregate" noise floor. For example, to have an overall noise floor (interference plus the receiver's thermal noise) of only 0.5 dB above the receiver's thermal noise, the interference level needs to be about 10 dB lower than the thermal noise. For the analysis data shown in Figure 1.18, a de-sensitization "target" of 0.5 dB is selected, thus the de-sensitization offset is 10 dB. Although radio designers typically assume 3 dB as a significant and measurable impact for communications systems, a 0.5 dB de-sensitization target was selected to encompass satellite receivers such as GPS (which have very stringent noise requirements) and to address the probable existence of multiple interference sources, including digital buses such as PCI Express (PCIe), which could have up to 16 lanes.

To illustrate the calculation, take 802.11g as an example. Starting with the thermal noise density, which is –173.8 dBm/Hz, we use the appropriate receiver bandwidth of 20 MHz for an 802.11g (as shown in Figure 1.19) to calculate the integrated noise floor for the receiver, and we get –100.8 dBm. The noise figure is typically 5 dB for an 802.11g radio. Adding the noise figure gives the overall noise floor of the receiver (not shown in Figure 1.18), which is –95.8 dBm. With 0.5 dB de-sensitization, another 10 dB is subtracted from the overall noise floor to give the allowable interference level of –105.8 dBm.

Radio Type	Channel B/W
802.11b	22 MHz
802.11g	20 MHz
802.11a (5 GHz)	16.6 MHz
Bluetooth (2.4 GHz)	1 MHz
UWB (3–5 GHz)	500 MHz
GPS (1.55 GHz)	2 MHz
GSM (850/900 MHz)	200 KHz
EDGE (1.8/1.9 GHz)	200 KHz
UMTS (1.8/1.9 GHz)	5.00 MHz
CDMA (925 MHz)	1.25 MHz

Figure 1.19: Receiver bandwidths.

It is interesting to note that all radios listed above require very low levels of interference to assure their performance (≤ 100 dBm).

The method used in the above example simplifies the analysis by reducing the complexity in dealing with the variation in sensitivity and SNR. It establishes a reference noise level to conveniently study the interference problem. For example, if one knows the aggressor's power level, the isolation requirement (between the aggressor and victim radio) can be easily calculated. From this level, one can calculate the SNR for the system and estimate its performance in terms of bit error ratio (BER), throughput, or range. If the actual noise is above the allowable interference level, the difference then becomes the reduction in SNR, which then can be used in system impact analysis.

Figure 1.19 shows the wireless receiver bandwidths of the common protocols. The amount of noise that the receiver sees will be dependent upon the bandwidth. The wider the bandwidth, the more platform noise the receiver will integrate. Figure 1.20 is a comparison of measured noise from two different digital devices in some well-known radio bands using a bandwidth of 1 MHz. As we can see, the measured levels of platform noise in the wireless bands are significantly higher than the allowable interference levels shown in

Radio Type	Center Frequency	Device 1 (dBm)	Device 2 (dBm)
802.11b	2.45 GHz	−65	−70
802.11g	2.45 GHz	−65	−70
802.11a(High)	5.8 GHz	−69	−63
802.11a(Mid)	5.3 GHz	−77	−76
802.11a(Low)	5.2 GHz	−67	−78
Bluetooth(2.4 GHz)	2.45 GHz	−65	−70
UWB(3−5 GHz)	4 GHz	−63	−54
GPS(1.575 GHz)	1.575 GHz	−58	−36
GSM(850 MHz)	880 MHz	−43	−41
GSM(900 MHz)	942 MHz	−41	−42
EDGE(1.8 GHz)	1842 MHz	−41	−39
EDGE(1.9 GHz)	1960 MHz	−68	−63
UMTS(1.8 GHz)	1842 MHz	−41	−39
UMTS(1.9 GHz)	1960 MHz	−68	−63

Figure 1.20: Measured platform noise in radio channels.

Figure 1.18. Before doing a direct comparison, it is important to correct for the appropriate bandwidths of the specific radio. The platform impact will need to be modified accordingly. For example, if the receiver bandwidth is 2 MHz, then the platform impact is raised by 3 dB. If the receiver bandwidth is 500 kHz, then the platform impact is lowered by 3 dB.

1.3.1.5 Analysis of an Interference Source

In the previous example, we can see that the aggregate noise from a digital device can be very substantial compared to wireless sensitivity requirements. We will now look at a specific single source within a device, namely, a 5 Gb/s high-speed interconnect bus. Figure 1.21 shows noise measurements for such an interconnect.

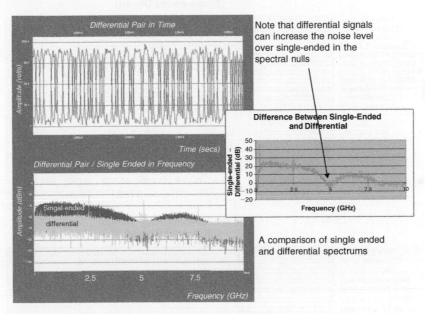

Figure 1.21: 5 Gb/s I/O bus: a comparison of single-ended and differential signals.

Differential signaling has become very common in high-performance digital devices such as personal computers. There are several reasons why this is the case, including improved EMI, improved signaling robustness to manufacturing variations, not to mention a faster, higher-performance bus under much the same cost basis. In order to access the impact from such interconnects to radio communication systems and determine if this differential signaling is beneficial to wireless interference, we will look at both signal-ended and differential responses. Figure 1.21 shows the resulting spectrum for both the single-ended case and the differential case. As we can see, at some frequencies there can be a significant benefit to using differential over single-ended. However, it is important to note that as the signal rate nulls, you can see higher noise levels for the differential signal than you can for a single-ended signal. It is therefore important to cover both cases as part of a complete analysis. In Figure 1.22, required isolation levels for our 5 Gb/s interconnect example are shown for both single-ended and differential signals.

Knowing the isolation requirements, we can perform impact analysis for the radios. Obviously, how much impact any interference source has on the radios depends on

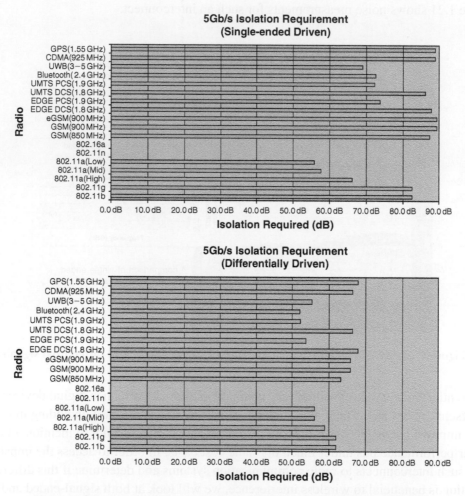

Figure 1.22: Required design isolation levels against radio thresholds.

implementation. We will first discuss the platform-level impact, then we'll follow with two case studies for module level and silicon level impact. Lower isolation means reduced receiver range for a given throughput or reduced throughput at a constant range. The table in Figure 1.23 illustrates the impact, where the isolation deficit (or additional isolation required) is related to wireless radio operating range reduction.

Interference Level	Operating Range Reduction for Constant Throughput
0 dB	0%
3 dB	19%
5 dB	30%
10 dB	50%
20 dB	75%

Figure 1.23: Range/throughput impact from interference.

For example, if an additional isolation of 10 dB is required (or 10 dB deficit in isolation), the range reduction will be 50%. This is true for all radios. The given table is for reference purposes and assumes a typical "indoor" environment. Actual range reduction depends on environment and may be slightly different.

The effect of platform isolation is somewhat difficult to analyze because each system is unique. Many parameters can vary, such as antenna location, radio module location, interferer location, high-speed bus routing, and so on. From one system to another, the impact can be quite different. Nevertheless, in a given design, the isolation requirements as shown in Figure 1.22 should be the design target.

Now that we have the isolation requirement from our analysis, we'll now consider what isolation is achievable given a known form factor or device implementation. In the following pages, we'll consider an industry standard module as our targeted form factor.

1.3.1.6 Isolation by Design: What's Achievable?

The PCIe Mini Card is chosen as an example for module-level impact analysis. A PCIe Mini Card is a 30 mm by 56 mm industry-standard form factor card that can be plugged into a slot on a notebook motherboard. It supports PCIe and USB 2.0 interfaces. Its primary function is to support wireless NIC designs. The test board is shown in Figure 1.24.

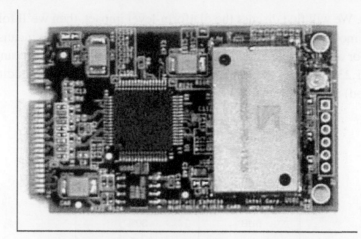

Figure 1.24: PCIe Mini Card sample.

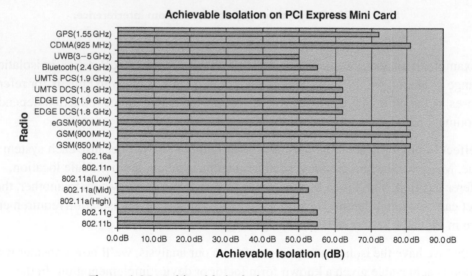

Figure 1.25: Achievable isolation on PCIe Mini Card.

A series of experiments were conducted on a PCIe Mini Card to determine its achievable isolation. Figure 1.25 shows the measured results from the analysis. The isolation was based on single-ended measurement.

Figures 1.26 and 1.27 show the additional isolation requirements for single-ended and differential signaling, respectively. This set of numbers is obtained by comparing the

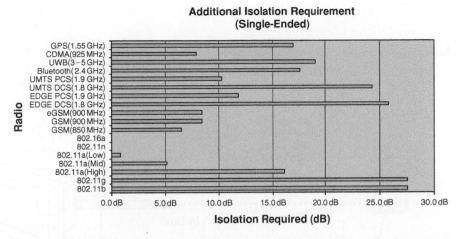

Figure 1.26: Additional isolation requirement (PCIe Mini Card, single-ended).

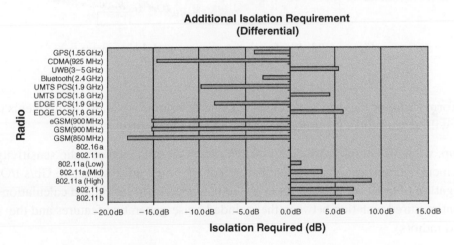

Figure 1.27: Additional isolation requirement (PCIe Mini Card, differential).

achievable isolation with the isolation requirement discussed in Figure 1.22. In some radio bands, as much as 25 dB of additional isolation may be required over and above what good design practice can achieve. In a multi-radio platform, these requirements may well be quite difficult to achieve. It's one thing to be able to design a bounded solution for a restricted range of frequencies; it's quite another to have to do the same over a range of radio bands that may encompass the full frequency range from 2 GHz to 4 GHz. That said, this is a good approach for designers to take for any new design. Perform an initial survey

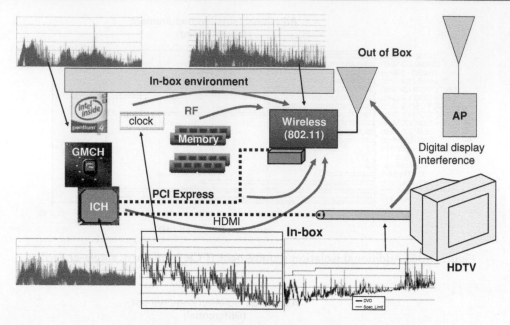

Figure 1.28: Platform components and their spectra.

of the form factor of the device under design, and determine isolation requirements that the design will have to meet to ensure the specified radio performance.

To recap, we have completed an analysis including determination of radio sensitivity requirements, identification and quantification of a specific noise source (5 Gb/s I/O bus), investigation of achievable isolation in a given targeted form factor, and calculation of the resultant required isolation between the included noise-generating features and the targeted radio(s).

We will now turn our attention back to a complete device or "platform."

1.4 Platform RF Snapshots

Figure 1.28 is a sketch of the components that compose the inside of the platform. Each device and function has its own distinctive spectrum and its own inherent impact to the wireless bands.

Figures 1.29 to 1.33 show a variety of wideband measurements of platform-generated noise. The first, Figure 1.29, is an example typical of noise measured during FCC testing

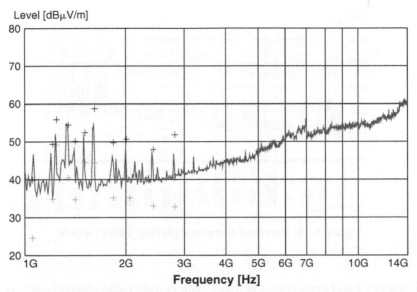

Figure 1.29: Measuring platform noise 1 GHz to 14 GHz at 3 m.

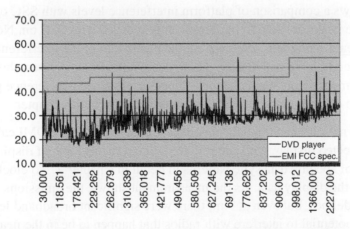

Figure 1.30: FCC 3 m measurement of DVD player emissions.

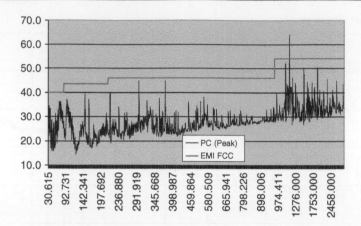

Figure 1.31: Personal computer playing a DVD movie.

across a frequency band encompassing many widely used radio channels. As we can see, there is significant noise across the GSM, GPS, and 802.11 bands.

Figure 1.30 shows the FCC result for a typical commercial entertainment device—a DVD player. Obviously, in addition to personal computers, more highly specialized devices produce complex spectrums that can become sources of interference to radios that are in close proximity. In fact, one of the authors has a wireless phone in his kitchen that kicks his notebook off the Internet whenever a call comes in. Microwave ovens have been known for some time to be sources of interference for 802.11b/g band devices.

Figure 1.32 shows a comparison of platform interference levels with SSC on and SSC off. The interference peaks are seen to be significantly reduced with SSC on. Note that while the measured peaks using SSC are lowered, the bandwidth of the interference peak is widened considerably. We will discuss this in detail later. In most cases, this will cause the interference to impact more radio channels, albeit at a reduced interference potential due to the specific time domain nature of SSC, as covered earlier in this chapter.

Figure 1.33 is a measurement of a display data stream over a 3 m HDMI cable. The data consists of a selected display sync pattern. We will cover this topic of display patterns and the impact of specific data sync patterns in detail in Chapter 2. No clock was forwarded with the display data. As the measurement shows, the emissions were well within FCC guidelines, but even at 3 m, there were significant broadband levels that could have the potential to interfere with radios that happen to be in the near vicinity of the source.

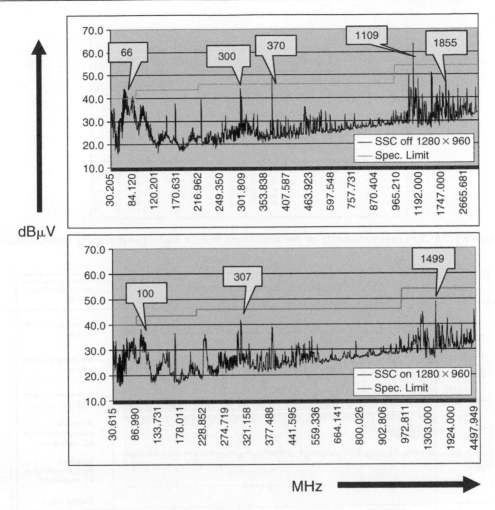

Figure 1.32: Impact of SSC in reducing peak amplitudes.

Figure 1.34 is a measurement of the platform noise as seen by the radio antenna located in the LCD of a notebook. Note the density of the spectrum as seen by the antenna. The overall background noise essentially fills the GSM bands and presents a difficult problem to the radio designer looking for high sensitivity and system performance. Figure 1.35 shows a similar measurement for the WiMax band.

In Figure 1.36, we show a method for bench top measurements that provide a fast means of characterizing the near-field interference potential of printed circuit boards,

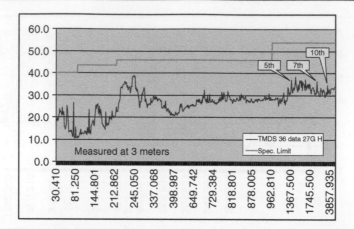

Figure 1.33: 3 m measurements of the emissions from a high-speed display interconnect.

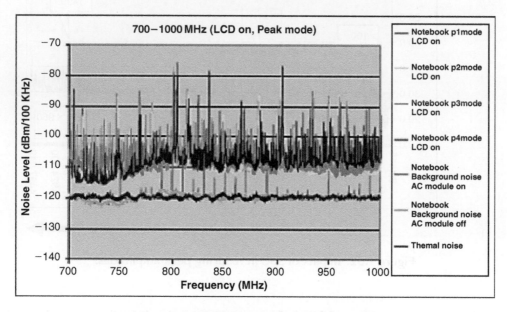

Figure 1.34: Platform noise in GSM bands.

components, or systems. The aperture of the horn antenna is approximately equivalent to the aperture of the motherboard of the platform, therefore we can be reasonably confident that we will be measuring only the emissions from the source and not the laboratory background. This particular horn antenna has a broadband frequency response that covers almost the entire set of wireless bands of interest.

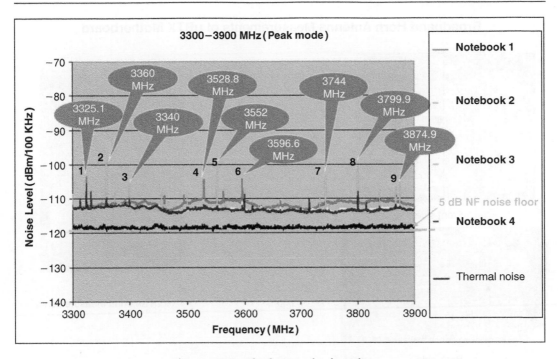

Figure 1.35: Platform noise in WiMax.

The interesting result of these measurements, as shown in Figures 1.37 to 1.40, indicates that the major source of the measured broadband emissions does not originate from the PCB trace structure, but rather from the areas associated with point sources (such as the heat sinks attached to major VLSI devices). We will return to this result in Chapter 7, when we consider analytical models for near-field radiators and the measurements that support the models and predictive processes.

The SinX/X curve shown in Figure 1.37 is meant to show the shape of a typical PRBS signal. When measuring a broadband signal in a real system, there will be many sources present and the clear shape of the PRBS signature may not be readily apparent. In Figure 1.37, we can see the shape of the broadband curve somewhat masked by the dominant narrowband spikes.

While the spikes may seem to be our primary concern (and they would be if we were concerned only with EMI/EMC), the broadband interference can be a larger problem when considering radio performance.

Broadband Horn Antenna Measurements of uBTX Motherboard

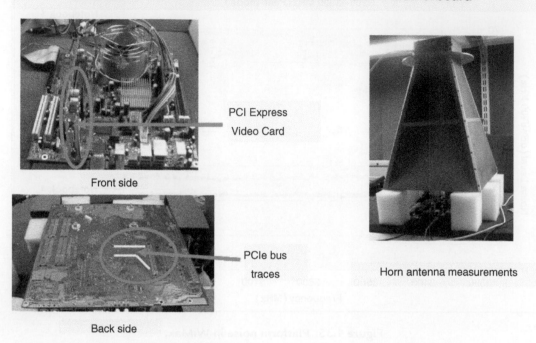

Figure 1.36: Measuring platform broadband emissions from PCBs.

Maximum emissions from the **front side** of the board when the board is running
3dmark2001se Benchmark (medium intensity) 3D graphics

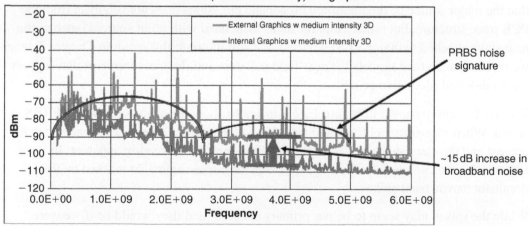

Figure 1.37: Measuring the platform noise on the component (front) side of a PCB.

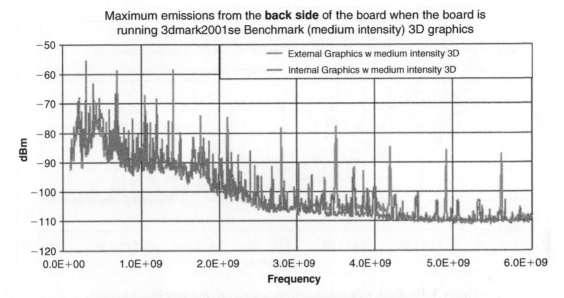

Figure 1.38: Measuring platform noise on the back side of a PCB.

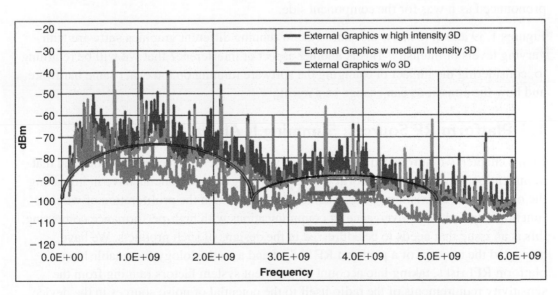

Figure 1.39: Front-side measurements with varying graphics intensity.

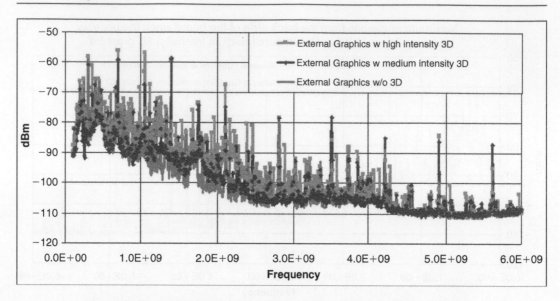

Figure 1.40: Back-side measurements with varying graphics intensity.

Note that the side of the PCB opposite the components shows emissions 20 dB down from the component side of the platform. Also note that the broadband noise signature is not as pronounced as it was for the component side.

Figures 1.39 and 1.40 show the same system running different graphics software with varying levels of intensity. This is another aspect of interference that we will be returning to, considering the impact of changing the software running on the exact same hardware and how the emissions can change as a result.

1.5 Platform RF Sources: Summing It All Up

In this chapter, we introduced the concept of wireless interference and the potential that nonintentional radiators in the form of high-speed digital components have in impacting the operation of coexisting wireless communications. With the proliferation of wireless functionality and the convergence of communications with high-performance computing, this is an issue that needs to be addressed in the designs of such products. We have presented the concept of a platform RF budget and a methodology for establishing platform RFI risks, taking into account all pertinent system factors ranging from the sensitivity requirements of the radio itself to the potential of noise sources in the device interfering with the integrated radios. In the concluding section of this chapter, we presented significant evidence that the issue of platform-generated RF noise is not something that exists only in the minds of a few individuals, but is indeed a real risk that must be addressed. In the following chapters, we will address in detail the potential noise sources and specific techniques that can be used to identify and quantify these sources within handheld, desktop, and mobile devices. In Chapters 8 and 9, we will present potential mitigation techniques that can be applied once a comprehensive understanding of the noise has been completed.

CHAPTER 2

The Structure of Signals

2.1 Time Has Little To Do with Infinity and Jelly Doughnuts

It is a striking fact that in the history and development of EMC/EMI, it has been a constant movement downward—from large-scale system design and mitigation to cables, to PCB, to the packages, and now to the silicon itself. Ultimately, this is where mitigation of interference should begin, at the silicon. It is in the silicon that the majority of signal creation occurs. We need to be concerned with the system signals—the movement and organization of electrical charge. That is the purpose of this book. We will explore package and silicon electromagnetics through direct measurement, analytical development, and theoretical considerations. We will attempt to show that a great deal of RF interference can be stopped entirely or lessened considerably right at the start in the silicon.

Two key measurement techniques that we will describe in detail are narrowband near-field scanning, where a fine-scale electric or magnetic field probe is stepped across the surface of the package or the silicon, and the energy distribution of the DUT is mapped; and broadband measurement using the VLSI GTEM, where a test board with a single IC is measured for broadband emissions from 100 MHz to 6 GHz. These two techniques are complementary in that once we have captured the broadband character of a device, we can then use the near-field scanner to investigate more closely the nature of the emissions structure inside the device at a set of predetermined frequencies. We will show just such an example later in the book, where a system clock device for PCs was measured using the GTEM, further investigation (with the NFS revealing point source attributes and the application of analytical techniques on the NFS scans) that led to significant improvement, 30 dB, in the RF interference profile of the device. Interestingly, the analytical approach is not as deep and difficult as one may think. It has been found that point sources, using the elementary dipole equations, can be applied quite profitably to describe complex radiation surfaces that approximate the real thing seen at the surface of silicon devices. These results follow directly from direct observation. This simplification gives us a 100X to 200X speed-up in computational time as compared to more exact 3D field solver methods, while still maintaining reasonable accuracy.

We will also show how one can use these measurements as guides to developing analytical methods to allow a designer to investigate possible functional floor planning early in the silicon and package design cycle. The work presented here, at heart, is meant to develop the techniques that allow us to design and build devices with a reasonably understood EMI/RFI profile long before they are ever actually physically placed in a larger system. Also, the intent is to help the gentle readers learn to investigate and develop techniques of their own.

2.2 Platform Signals: Preliminaries in Spectral Analysis

We begin this chapter by examining the structure of the signals that platforms generate. We will discuss rise times, fall times, and duty cycle and the impact on the spectra of signals. We will show that symmetry and asymmetry of the signal shapes have significant effects. Primarily, we will look at clocks, data, and other dominant symbol structures, such as those constructed for digital displays. All of these signals are present in the main system ICs. Understanding their spectral content will give us insight into how they can couple to incidental antennas such as heat sinks and power planes.

Figure 2.1(a) shows the structure of a typical clock signal; (b) shows the phase variation of the signal; (c) is the first derivative of the signal with respect to time, and (d) is the phase variation of the derivative. We wish to investigate the various components of the signal structure, such as duty cycle—how the harmonics vary with time. We need to know how the spectrum changes and rearranges itself when the symmetry of the signal is changed. The important pieces of the signal are shown: the edge rates, both rising and falling, will determine how far up in frequency we will need to work; the amplitude of course; and the period of the signal. The period is the amount of time that the signal is sending energy. If it is on all the time, we have a DC signal; as we start reducing the on-time, we eventually approach an impulse that mathematically means that the signal is on for an infinitesimally short time. As we will see, when we approach this type of signal, the upper frequency limit becomes quite high. Signals such as clocks will typically have a 50% period, or duty cycle, such as that shown in Figure 2.1.

We will use Equations 2.1 and 2.2 to perform these investigations. The signal is broken into three pieces for analysis: the first piece describes the rise time; the second piece

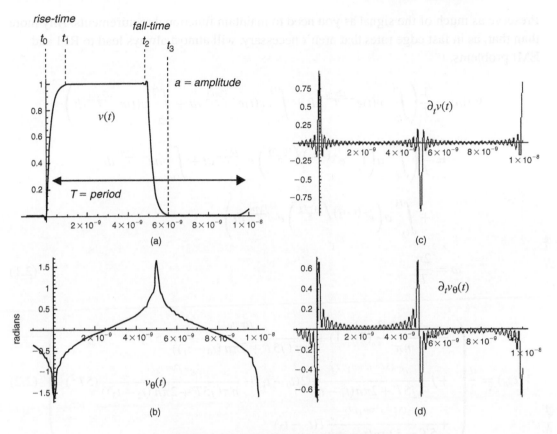

Figure 2.1: Structure of a clock signal.

concerns the steady-state portion of the signal; and the third piece describes the fall time. These three pieces compose the Fourier Transform of the signal under study. We are taking a time domain description of the signal and transforming it to the frequency domain. We are doing this to be in a better position to determine how the signal may cause interference in our radio bands. By moving into the frequency domain, we can then plan and adjust our signals to produce less interference in the radios.

The second equation given is the solution to Equation 2.1. Note that it is a function of rise time, fall time, period, and harmonic number, *n*. If we could take an infinite number of harmonics, we would exactly reproduce the signal when we transform back into the time domain. In practice, we don't need to do that; we can show that 15 to 20 harmonics will pretty much guarantee that the signal integrity is preserved. This is an important notion.

Preserve as much of the signal as you need to maintain functional requirements. Any more than that, as in fast edge rates that aren't necessary, will almost always lead to RFI and EMI problems.

$$V(\omega) = \frac{1}{T}\left(\int_{t_0}^{t_1} v(t)e^{-\frac{j2\pi nt}{T}}\,dt + \int_{t_1}^{t_2} v(t)e^{-\frac{j2\pi nt}{T}}\,dt + \int_{t_2}^{t_3} v(t)e^{-\frac{j2\pi nt}{T}}\,dt \right)$$

$$= \frac{1}{T}\left(\int_{t_0}^{t_1} a\left(1 - e^{-(t-t_0)/\frac{t_1-t_0}{5}}\right)e^{-\frac{j2\pi nt}{T}}\,dt + \int_{t_1}^{t_2} ae^{-\frac{j2\pi nt}{T}}\,dt \right.$$

$$\left. + \int_{t_2}^{t_3} a\left(e^{-(t-t_3)/\frac{t_3-t_2}{5}}\right)e^{-\frac{j2\pi nt}{T}}\,dt \right)$$

$$\omega = \frac{2\pi}{T} \tag{2.1}$$

$$V(\omega_n) = \frac{a}{2}\left(\begin{array}{l} -j\dfrac{e^{-j2\pi nt_1/T}}{n\pi}(T-1) + \dfrac{e^{-j2\pi nt_0/T}}{n\pi\,(j5T + 2\pi n\,(t_0 - t_1))}(5T) \\[2mm] +j\dfrac{e^{-(5T+j2\pi nt_1/T)}}{-j5T + 2\pi n(t_1 - t_0)}2(t_0 - t_1) - \dfrac{e^{-j2\pi nt_2/T}}{n\pi(j5T + 2\pi n\,(t_2 - t_3))}(5T^2) \\[2mm] +\dfrac{e^{-(5T+j2\pi nt_3/T)}}{5T - j2\pi n(t_2 - t_3)}T(t_2 - t_3) \end{array} \right) \tag{2.2}$$

Equation 2.1 will deliver the individual harmonics, both even and odd, which fully and completely describe the signal. The equation as given incorporates the effects of asymmetry between rise and fall edge rates and duty cycle.

2.3 Harmonic Variation with Time

Figure 2.2 shows the relation between time and harmonic number. The plot is a continuous surface, whereas a measured signal will be composed of a discrete set of harmonics. We show this surface as a heuristic toward gaining a better understanding of signal structures. One observation right off the bat is to see that the first harmonic is the strongest in terms of amplitude. The first harmonic energy will vary slowly with time; higher harmonics, though much lower in amplitude, will vary more quickly in time and therefore could appear as a strongly varying interference source in a receiver. The slowly varying harmonic will be

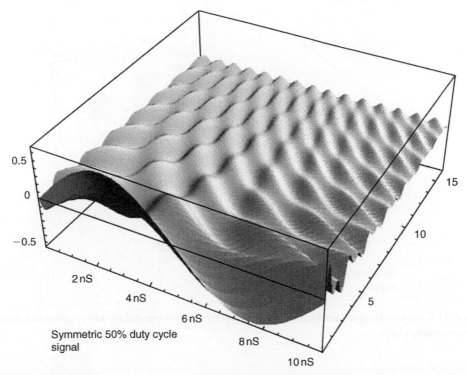

Figure 2.2: Constructing a continuous surface relating time variation with harmonic number for a symmetric clock.

more amenable to mitigation techniques that reside in the radio software; the faster varying harmonics will be harder to track and mitigate.

Sometimes showing a discrete set of points as a continuous surface allows the viewer to recognize or discern patterns in the data that otherwise might not be apparent.

Figure 2.3 zooms in on the first five harmonics shown in Figure 2.2. Figure 2.4 zooms in further on harmonics 3, 4, and 5. These two plots clearly show the time variation associated with each harmonic and how that variation can smoothly transition between the discrete harmonics if we could allow continuous harmonic values.

Figure 2.5 shows the discrete harmonic variation and Figure 2.6 is an extension of this and is another example of extending a set of variables to a continuous set. In this example, we vary rise time and harmonic number. We hold the fall time constant at 100 pS. The purpose here is to show the impact of edge rate asymmetry in the set of even and odd harmonic

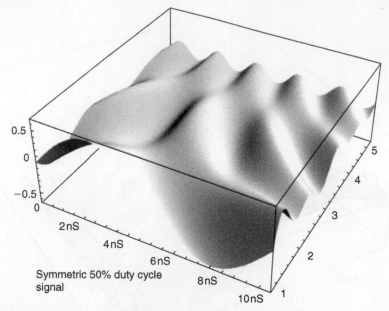

Figure 2.3: Constructing a continuous surface relating time variation with harmonic number for a symmetric clock.

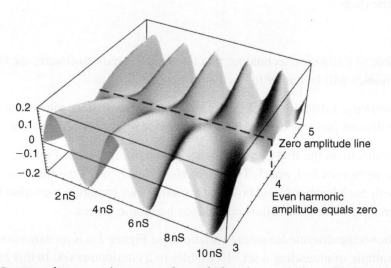

Figure 2.4: Constructing a continuous surface relating time variation with harmonic number for a symmetric clock signal (3rd to 5th harmonics).

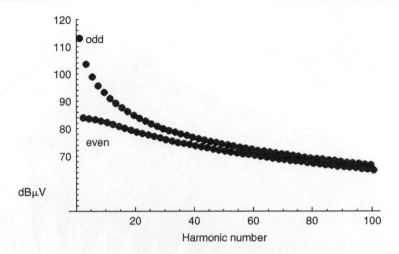

Figure 2.5: Even and odd harmonics as a function of rise and fall time asymmetries.

amplitudes. As Figure 2.6 shows, there is a significant increase in the even harmonic amplitudes. Again, for the lower harmonics, this effect is not so pronounced. In the higher harmonics, the even and odd harmonics become equal, and, at some point in the spectrum for asymmetric signals, the even harmonics will be of higher amplitude than the odd harmonics.

Signal asymmetries are one of the most common causes of interference. Most people assume that a trapezoidal clock signal will only have odd harmonics. In most real systems this is just not the case, and the even harmonics must be accounted for. How they arise is an important process to understand.

These surface images reveal the inner structural beauty of this type of analysis. One can see emerging symmetries and how these symmetries slowly change and morph. Overall, the beauty of the mathematics comes through.

2.4 Variation with Duty Cycle

Figures 2.7 to 2.9 show the edge rate asymmetry in closer detail. Figure 2.7 is the symmetric clock signal. In Figure 2.8, the fall time is changed from 500 pS to 550 pS. As we can see, even harmonics appear, and the total power delivered increases by a few percent. In Figure 2.9, the asymmetry has been increased to 1000 pS in the fall time and the even harmonics have increased correspondingly, also increasing power delivered by 5%.

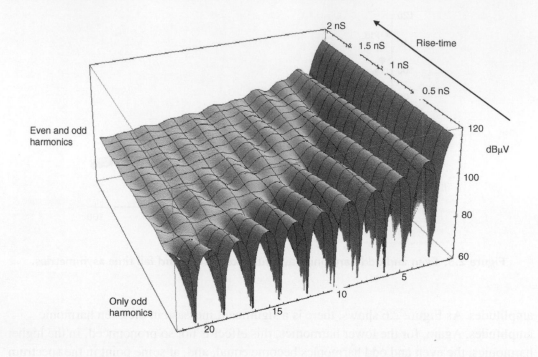

Figure 2.6: Edge asymmetries give rise to even harmonics.

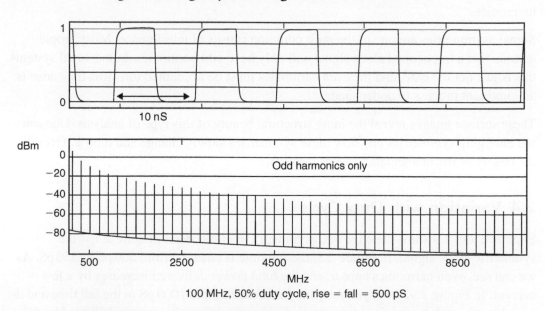

Figure 2.7: A symmetric trapezoid: Only odd harmonics are present.

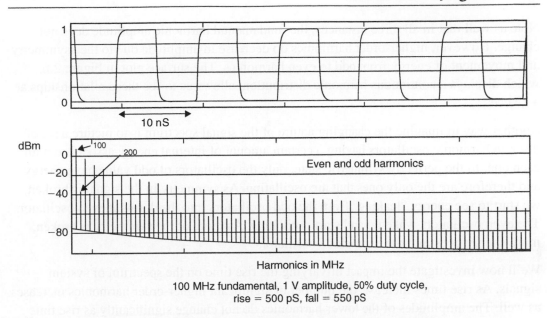

Figure 2.8: Asymmetry in the signal creates even harmonics.

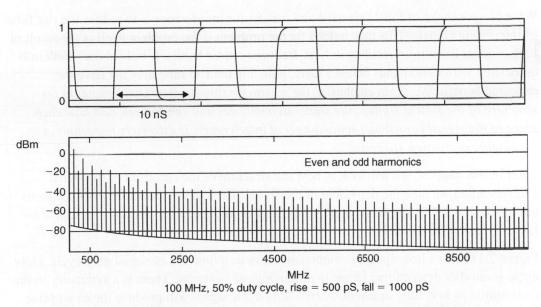

Figure 2.9: Increasing the asymmetry increases the even harmonic amplitudes.

Note as well that in all three instances, the fundamental harmonic amplitude does not change. However, higher-order harmonics do decrease in amplitude due to the asymmetry and movement of energy from odd to even harmonics. The surface plot in Figure 2.6, which displays a continuous harmonic distribution, allows us to see these relationships at a glance.

Another way to imagine the changing nature of the signal spectrum is to picture a set of discrete harmonic oscillators having a certain amount of internal energy associated with each one. In the perfectly symmetric state, only the oscillators of odd value have energy and therefore are the only ones that are oscillating. As symmetry in the signal is broken, we start to see energy "leaking" from the odd oscillators into the dormant even oscillators. The even oscillators wake up and start oscillating. As more of the symmetry is broken, more energy goes into the even mode oscillators.

We'll now investigate the impact of varying the rise time on the spectrum of system signals. As rise times increase, the number of significant higher-order harmonics increases as well. The amplitudes of the lower harmonics do not change significantly as rise time increases. The following figures illustrate this. In Figure 2.10, note that the 3rd harmonic amplitude changes by only 2 dB while the 19th harmonic increases by 14 dB. Also note that the spectrum depends on the edge rate of the fastest edge of the signal.

What can we make of this? One tried and true fix for interference is to reduce the rise time of interfering signals. This may indeed fix the problem if the problem itself is the result of higher-order harmonics residing in high-frequency radio bands. If instead the problem is associated with radio bands below 1 GHz, reducing the rise time may not have the expected benefit. We again emphasize the reason for this chapter: *Understanding the structure of the system signals, the signal asymmetries and symmetries, and how each piece of the signal contributes some aspect of interference, is extremely important in developing mitigation strategies.*

Later in this chapter, we will look at how the structure of the spectrum of the signal changes when we take the derivative of the signal. We will see that the lower harmonics are much reduced in amplitude and that the higher-frequency harmonics do not have as fast a reduction in amplitude.

Figure 2.11 shows how spectral components vary as a function of signal duty cycle. Duty cycle is another determining factor in producing asymmetries. There is a symmetry to this distribution as is readily apparent. A 10% duty cycle signal will produce the exact same spectral components as a 90% duty cycle signal. This can be demonstrated by simply

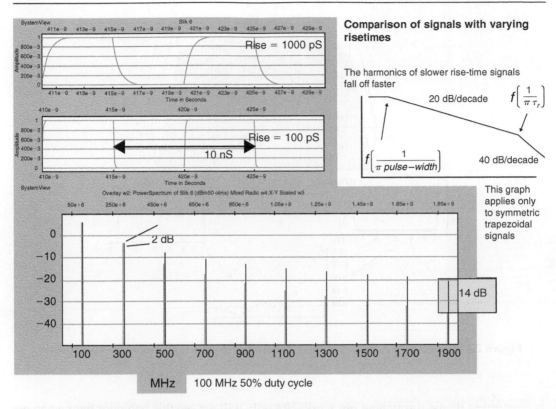

Figure 2.10: Rise time changes affect higher-order harmonics the most.

shifting the baseline, or the DC offset, of the two signals such that the 10% signal becomes a 90% duty cycle signal with negative amplitude. (Spectrum analyzers measure absolute value.)

Interestingly, when we compare the imaginary and real components, we can see that the imaginary components have strong lower harmonic symmetry while the real components have strong lower harmonic asymmetry. This is shown in Figures 2.12 and 2.13. Notice the overall symmetry of the harmonic distribution when we allow the harmonic number to become a continuous variable. If we plot the discrete set of harmonic components, we may not see this structure as readily. Throughout our work, we will return and use this approach to look for underlying structures.

Compare Figures 2.12 and 2.13, the real and imaginary distributions. Especially note the first harmonic variation in each and note that the two are out of phase by 90 degrees. This

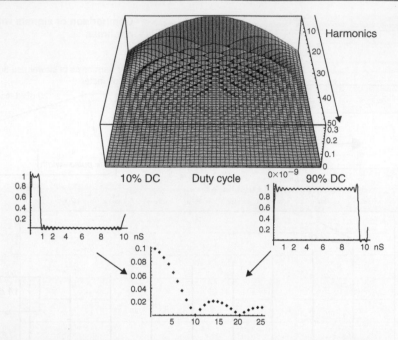

Figure 2.11: Spectrum variation with signal duty cycle; the real and imaginary spectra.

is a recurrent theme throughout our work. Not only will we see this behavior here when we consider signal structures, but we will also see it in Chapter 3, when we consider the radiated emissions of elementary point sources and observe that the relationship between some far-field and near-field components is out of phase by 90 degrees.

Something else to appreciate here is simply the beauty of the symmetry of the harmonic distribution. When we look at a sufficient number of harmonics, the symmetry becomes more apparent as does the overall harmonic structure.

We'll now look more closely at the effect of edge rates. Figure 2.14 will be our starting point.

In order to investigate very low duty cycles, we set the rise time at a very fast rate: 20 pS. This may seem fast now, but with expected process changes in silicon, 20 pS will become more common as a signal rise time. Note that a 100 MHz clock with 20 pS rise time can have significant energy all the way to 15 GHz and beyond. The signal structure determines how the harmonic energy will fall off with frequency. There are two important breakpoints in the envelope that describes the harmonic maximums. The first is a function of the

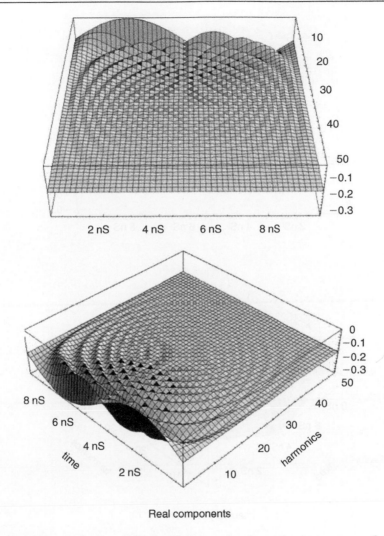

Figure 2.12: Spectrum variation with signal duty cycle; the real components of the spectrum.

pulse-width and occurs at a frequency equal to

$$\frac{1}{\pi \text{ pulse-width}}.$$

The harmonic amplitudes will fall off at a rate equal to 20 dB per decade of frequency. The second breakpoint is a function of the fastest edge rate, either the rise time or the fall time,

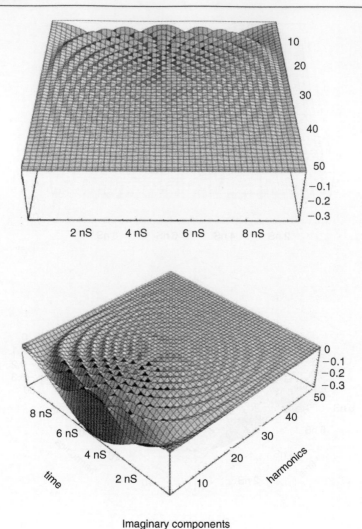

Figure 2.13: Spectrum variation with signal duty cycle; the imaginary components of the spectrum.

and occurs at a frequency equal to

$$\frac{1}{\pi \,\text{rise time}}.$$

The harmonic amplitudes will fall off at a rate equal to 40 dB per decade of frequency. Almost every EMI/EMC textbook ever printed repeats these breakpoint calculations, but

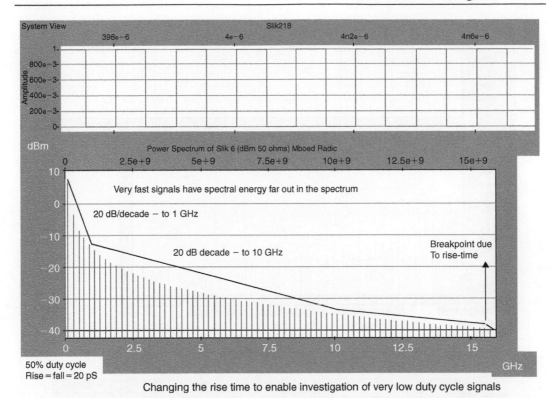

Changing the rise time to enable investigation of very low duty cycle signals

Figure 2.14: The faster the rise time, the greater the spectral content.

almost none will add that it applies only to the pure trapezoidal Fourier components and might not apply to a real signal driving a real physical channel. We will discuss this in more detail later in the book.

In Figure 2.15, we change the duty cycle from 50% to 40%. Note the change in the spectral envelope. There is no longer the 20 dB and 40 dB envelope of maximums as shown in Figure 2.14. Instead, we see some harmonics disappear entirely and others increase in amplitude. Figures 2.16 and 2.17 show the spectral impact of decreasing the duty cycle further. When the duty cycle is 10%, we start to see the envelope usually associated with impulses. Clear nulls are forming at $f_n = n/pw$. In Figure 2.18, with a 0.5% duty cycle, we see only a small roll-off in harmonic amplitude through 15 GHz, with an approximate 45 dB decrease in amplitude from a 50% duty cycle.

For most EMC applications, a very low duty cycle signal has a low associated probability of interference. There is a 50 dB difference between the 50% duty cycle signal and the

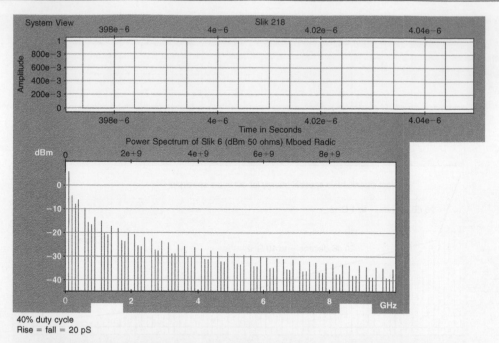

Figure 2.15: Changing the duty cycle changes the spectral distribution.

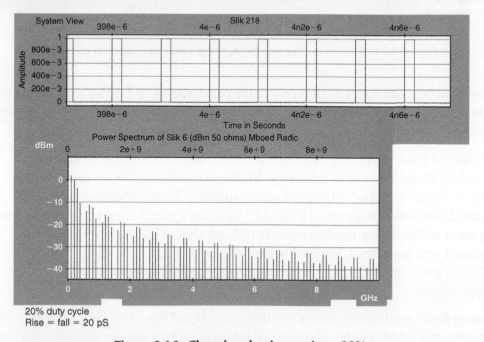

Figure 2.16: Changing the duty cycle to 20%.

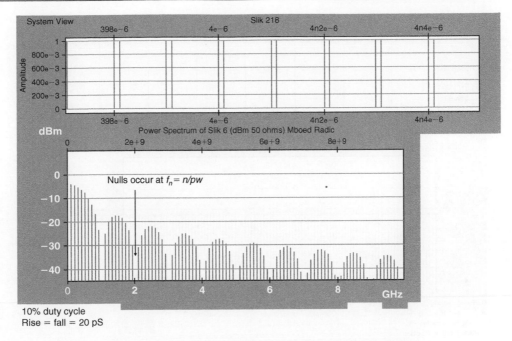

Figure 2.17: At 10%, spectral nulls are clearly forming.

0.5% signal. However, when we start worrying about radio interference potentials, this 50 dB difference may not be margin enough, especially since we're concerned with system sources and radio receivers that will be located extremely close to each other. As Figure 2.18 indicates, we may have a low duty cycle signal, but the harmonic content does not fall off in amplitude for tens of GHz, and the levels of these harmonics are well within radio receiver thresholds. Therefore, we have a problem that we do not have in EMC/EMI design.

Figure 2.19 shows the time variation of the harmonics for a 10% duty cycle pulse. Figure 2.20 shows a 300 pS pulse at a 100 MHz repetition rate. It is interesting in this plot that from low harmonic to high harmonic there is very little variation in amplitude, while the increasing time variation remains.

So what are we saying at the end of all this? Low duty cycle, high-frequency signals with fast rise times can produce interference across many radio bands simultaneously. However, even though a signal may be low in frequency, such as a 1 MHz clock or other timing type, the presence of a fast edge can produce energy that may interfere in radio bands.

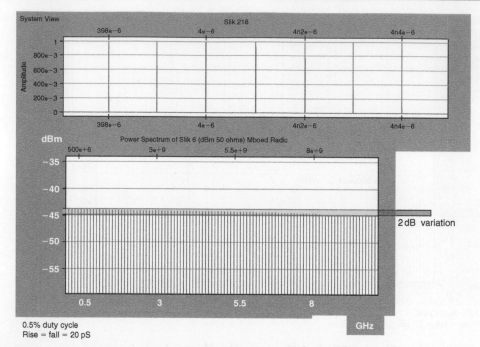

Figure 2.18: The maximum amplitude decreases, but envelope fall-off is approximately constant.

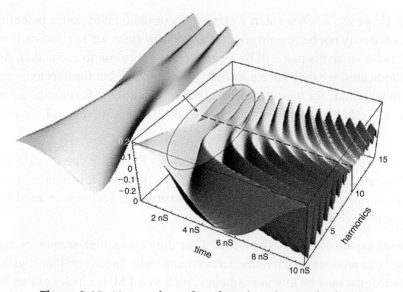

Figure 2.19: Harmonic surface for a low duty cycle signal.

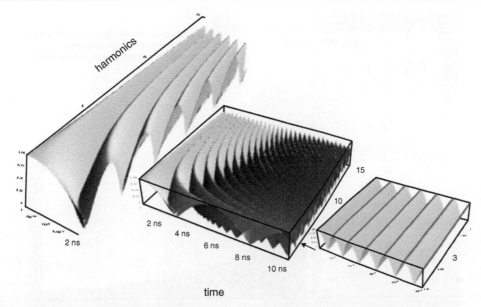

Figure 2.20: Narrow pulses produce harmonic components with very little amplitude variation.

Figure 2.21 shows how even a low duty cycle low-frequency signal can produce a harmonic rich environment. We show two instances: a 1% duty cycle 100 MHz signal, and a 1% duty cycle 1 MHz signal.

2.5 Nonrepetitive Signals

Figure 2.22 shows the resulting spectrum of a true impulse. Momentarily, an impulse can produce an energetic and broad spectrum roughly 100 dB down in average peak amplitude from the 50% duty cycle clock, and instantaneously down about 80 dB as seen at 100 MHz. Figure 2.23 shows the time behavior. In Figure 2.24, a step function is shown for both the spectrum response and for the time domain at 100 MHz. For the step function, the instantaneous value is only 25 dB down from the 50% clock. The average peak amplitude is 90 dB down. Figure 2.25 shows the time domain behavior of a step function as seen by a radio receiver with two different bandwidths: 4 MHz and 24 MHz. As we can see, the narrower the bandwidth, the longer the interference is at significant levels; for a wider receiver bandwidth, the amplitude is higher but the disturbance lasts a much shorter time. As these examples show, the platform designer must be aware of the spectral behavior of such signals as impulses and step functions, and how the spectrum interacts with the front end of a radio receiver.

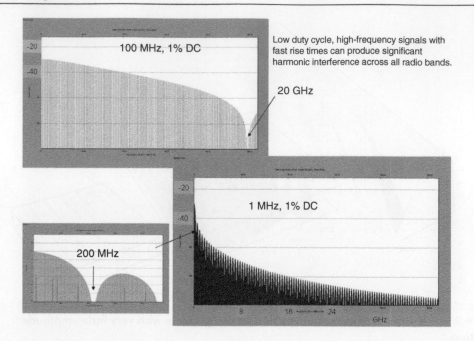

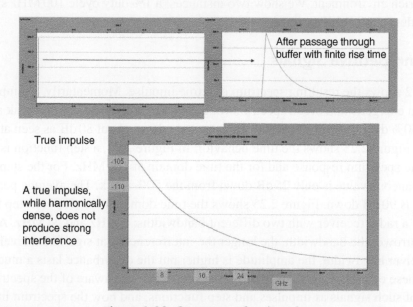

Figure 2.21: Low duty cycle repetitive signals.

Figure 2.22: Spectrum of an impulse.

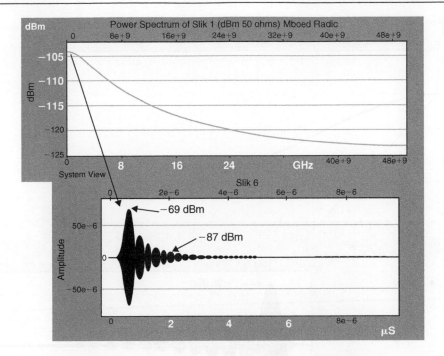

Figure 2.23: Time behavior of an impulse at 100 MHz.

In EMC/EMI work, we would tend to ignore these signals as not being typical contributors to system failure. We may spend some time considering low duty cycle signals that could have an impact on a quasi-peak detector as used in EMI receivers, but impulses and step functions typically do not charge the quasi-peak detectors to the extent that the signal crosses the threshold limits during testing. In RFI and radios, it may be that the impulse and the step function may occur at such a rate that they have an impact on the functioning of our radios. We need to understand what this impact looks like in terms of symptoms and spectral response so we can be in a better position to identify which piece of the platform is producing the interference.

2.6 Modifying a Clock

Figure 2.26 shows a 100 MHz clock that has had its spectrum spread out in the frequency domain. Spread spectrum is widely used to reduce EMI levels. The idea is that a modulating signal is applied to the clock signal, typically around 30 KHz and looking something like a triangle wave. The clock harmonics are then moving in frequency around the harmonic value by whatever value of spread is chosen, typically something around 0.5%. Since the

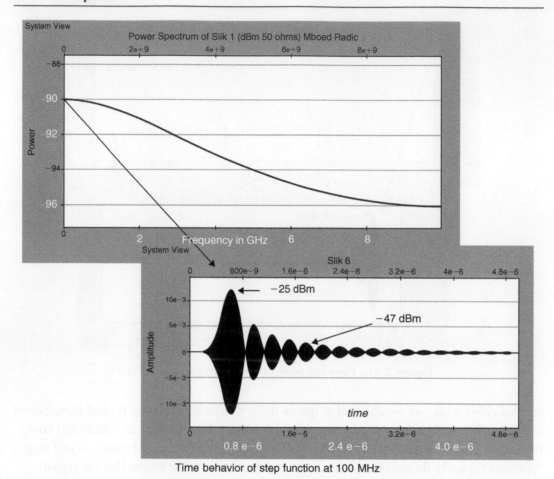

Figure 2.24: Spectrum of the step function.

same amount of energy is present, this means that the energy is spread among the frequencies centered on the harmonic with the result that the max peaks are lowered. As we can see, the effect of spread spectrum increases as harmonic number increases. The 500 MHz narrowband signal has been mathematically shifted down from the spread signal to better show the comparison.

So spreading a clock can reduce the amplitude, but we don't get something for nothing; we end up with a lower amplitude signal, but its effect is spread over a wider frequency. While this approach may have quite good EMC/EMI benefits, the RFI impact may actually be worse than if we had done nothing.

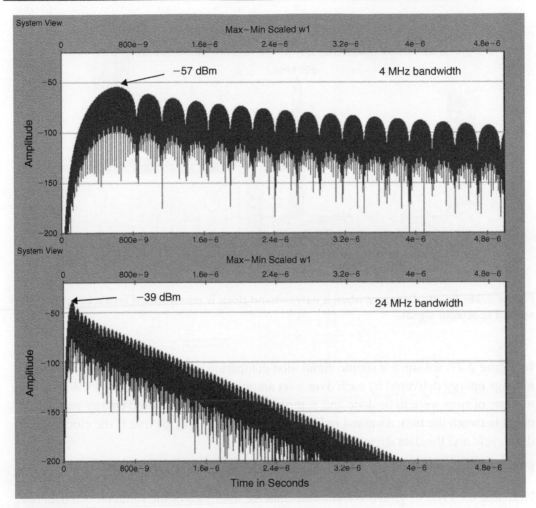

Figure 2.25: Spectrum of step function in the time domain in the WLAN band for two receiver bandwidths.

2.7 Random Data Sequences as Signals

Clocks are generally, and for good reason, considered the leading causes of interference in platform systems. They are generally 50% duty cycle signals and have fast edge rates for low jitter requirements. We will now look at two other types of signals found in platforms: data-type signals, which can be modeled by pseudorandom bit stream (PRBS) signals, and display symbol streams. We'll look at PRBS signals first.

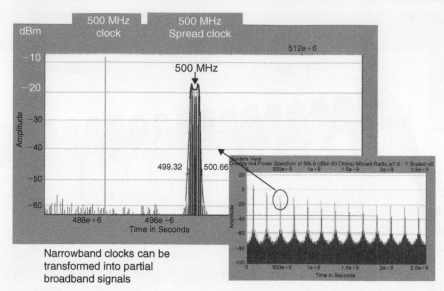

Figure 2.26: Spectral content when a narrowband clock is transformed into a broadband spread spectrum signal.

In Figure 2.27, we show a simple simulation comparing a clock and a data signal and the average energy delivered by each over a set amount of time. They differ by 1%. If a large number of runs were to be done and compared, the average delivered energy would be equal between the two. As noted in Figure 2.27, this will only be true if the clock is 50% duty cycle and the data stream is DC-balanced.

Figure 2.28 plots a typical PRBS signal and its corresponding spectrum.

Note that the PRBS signal produces nulls that are multiples of the fundamental data rate. In addition, when the underlying bit structure is not symmetric (whether due to edge or pulse-width jitter), there arises a narrowband spike at harmonics of the data rate. In other words, the nulls of PRBS signals are not nulls at all! We will return to a study of edge and pulse-width jitter at the end of this chapter.

The PRBS signal will produce the lowest general spectrum (leaving out the true impulse). This is opposite in impact from the clock signal, which produces the highest spectral peaks. However, the PRBS signal will be broadband while the clock will be narrowband. Again, we need to think about radio interference and not be too constrained by our previous EMI/EMC experience. The following plots show comparisons of PCIe signals to simulated PCIe data streams. The impact of equalization is shown.

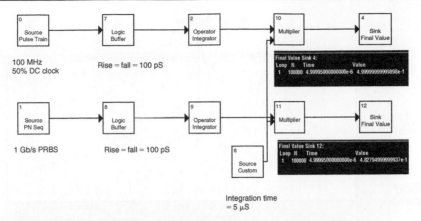

100 MHz
50% DC clock

Rise = fall = 100 pS

1 Gb/s PRBS

Rise = fall = 100 pS

Integration time
= 5 µS

A DC balanced PRBS signal delivers the same energy as a 50% duty cycle
clock (on average)

Figure 2.27: Comparing energy delivered for a clock and for a PRBS signal.

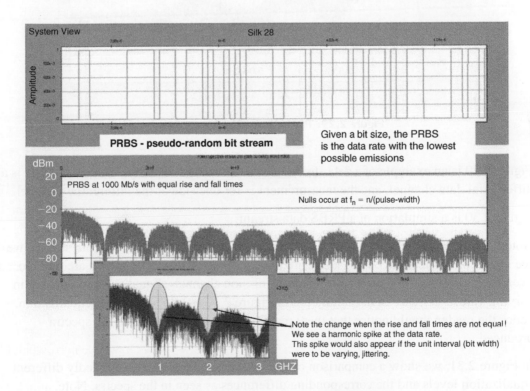

Figure 2.28: The spectrum of a PRBS signal.

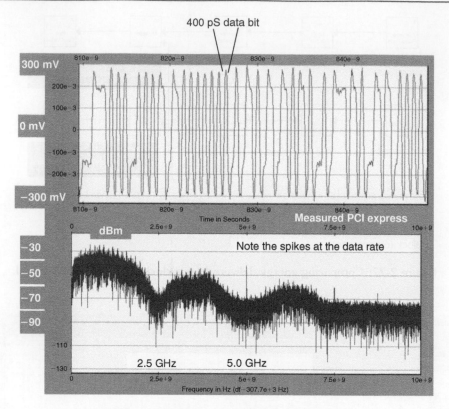

Figure 2.29: PCIe data stream, measured.

Figure 2.29 is an example of a data stream measured on a PCIe platform channel. PCIe is a differential data channel, and the measurement is for one side of the differential channel.

Figure 2.30 is a simulation of a PRBS data stream.

Note the difference between Figures 2.29 and 2.30. In the measured PCIe data stream, we see spikes on the leading edges. These are the equalization functions, which are there to overcome high-frequency effects of the transmission channels. Equalization can come in varying values in PCIe. In the simulation in Figure 2.30, we do not include equalization, hence the reader should notice that there really isn't much difference in the spectra produced.

In Figure 2.31, we show a comparison of simulated data streams with markedly different equalization levels and the corresponding differences as seen in the spectra. Note, however, that the difference is seen in the null spikes and not in the general level of the

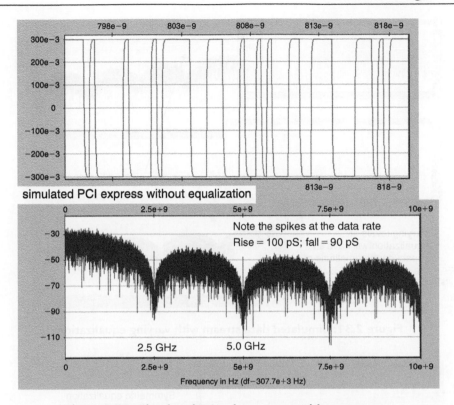

Figure 2.30: Simulated PCIe data stream with asymmetry.

spectra, and that these equalization functions are applied to only one edge of the data stream. Figure 2.32 shows the simulation streams when equalization is added to both edges and the spectrum effects when we vary the symmetry of the equalization.

So, we see again that asymmetry in any aspect of the signal structure introduces spectral components that would not be there if the signal was truly symmetric. Jitter in the pulse width, jitter on the edge rate, a rise time not equal to fall time, and equalization of the signal all can contribute in varying amounts to spectrum modification. As we become more familiar with the effects of each asymmetry, we will become more adept at recognizing where in the platform an interferer may be found.

Figure 2.33 shows one more aspect of the signal structure that we need to keep in mind when considering the impact of signal interference. In most instances, when signal spectra are compared, we take the spectra of signal sources as they would be seen at the input to a

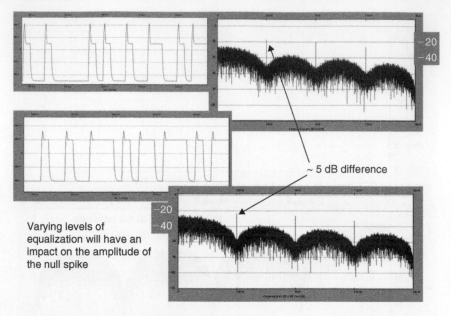

Varying levels of
equalization will have an
impact on the amplitude of
the null spike

~ 5 dB difference

Figure 2.31: Simulated data stream with varying equalization.

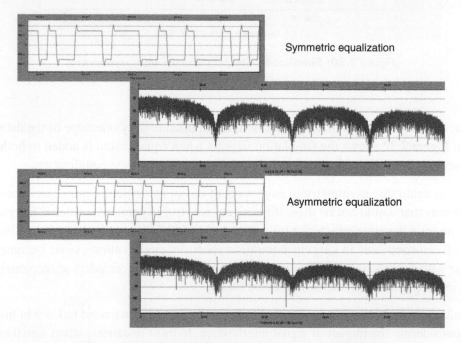

Symmetric equalization

Asymmetric equalization

Figure 2.32: Symmetric and asymmetric equalization.

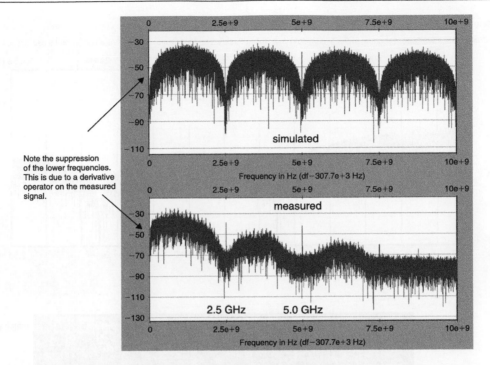

Figure 2.33: Applying the derivative operator to the data stream.

spectrum analyzer. The transmission channel is usually thought of as a coaxial cable, directly connecting the signal source to the analyzer. However, this is neither the channel nor the mechanism through which the interference is coupled to the radio input. The typical transmission paths will be either conducted or radiated. In both instances, the signal as seen by the radio will look like the differentiated signal, as shown in Figure 2.33.

Note that the effect of the time differentiation of the signal is to decrease the amplitude of the lower-order harmonics with little impact on the higher harmonics. In some cases, this effect can make our work easier. The lower harmonics, especially for clocks, have the greatest amount of energy. Our life is certainly easier if our mitigation efforts can be bandwidth limited.

Figure 2.34 shows an interesting way to convert a deterministic spectrum into a random spectrum. The idea is to use a scrambling circuit to take the clock and produce a spectrum much like a PRBS spectrum. In the example shown, the nulls are related to the higher-frequency clock used in the scrambling circuit—in this case, it is a 1 GHz clock. As shown in the plot, 25 dB reductions in harmonic peaks are achievable.

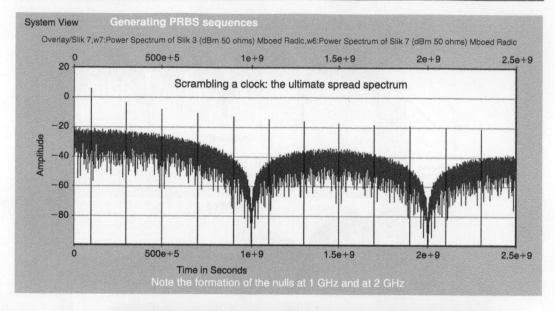

Figure 2.34: Scrambling a deterministic signal produces a PRBS spectrum.

2.8 Spectra of Display Symbols

By display symbols, we mean the following:

In the example in Figure 2.35, the clock can be thought of as a symbol with 5 consecutive bits set to "1" and 5 consecutive bits set to "0." The system display symbols will compose a finite set of complex signals. In most cases, each symbol will also be required to be DC-balanced, as is the symbol shown in Figure 2.35.

Figure 2.36 shows three arbitrarily constructed 10 bit symbols. Their time domain structure and their resulting spectra are shown. At some harmonics, there can be a 35 dB variation between harmonic peaks. It's not hard to imagine that the corresponding interference differences between symbols in any given symbol set can be quite large.

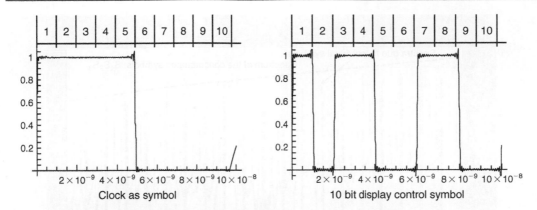

A symbol is a complex bit sequence usually employed in display systems for sync functions.

Figure 2.35: Display symbols as complex bit sequences. A 10 bit symbol is shown.

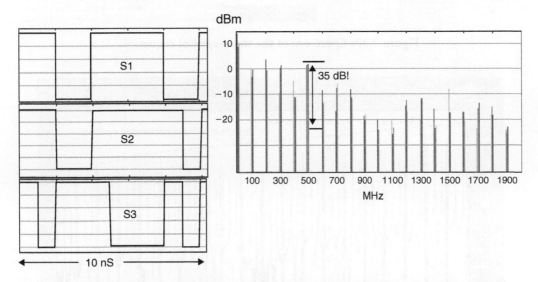

Figure 2.36: Comparison of spectra for arbitrary 10 bit symbols.

Figures 2.37 and 2.38 show a comparison of the spectra for the three symbols, with an additional example—the resulting spectrum when the three symbols are combined in a series. As we can see, the series spectrum has more harmonic peaks to it, but each peak is lower in amplitude than the envelope of the maximum peaks from each of the separate

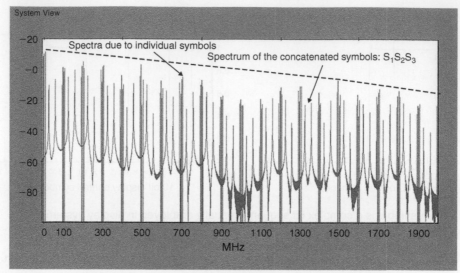

Combining the symbols in a particular order and repeating

$$S_{123} = S_{231} = S_{312}$$

Figure 2.37: Spectrum of the three-symbol sequence.

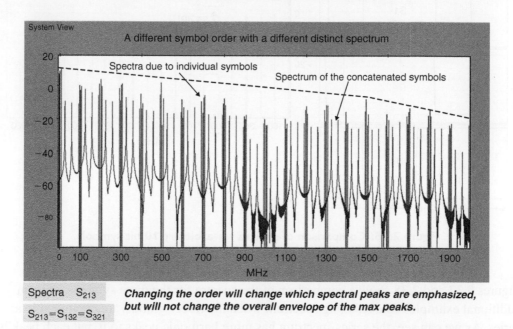

Spectra S_{213}

$$S_{213} = S_{132} = S_{321}$$

Changing the order will change which spectral peaks are emphasized, but will not change the overall envelope of the max peaks.

Figure 2.38: Spectrum of the three symbols in a different sequence.

symbol spectra. This is intuitively understood when one realizes that there are more symbols in which the energy can be assigned. In the limit, if the symbol set were to grow in the number of symbol elements with each symbol taking an equal amount of energy over a given time duration, the spectrum would become that of a PRBS spectrum. Additionally, as shown in Figure 2.38, if we rearrange the time sequence of the symbols, the harmonic energy redistributes.

This is an important observation for us. It provides one more tweaking knob for us in our quest to find the platform features we can adjust in order to assure superior wireless performance. We see that our symbols each produce spectra of their own, and that adding them to form a distinct time sequence produces a unique spectrum. We can therefore adjust how we construct these sequences in order to customize where in the spectrum their harmonic energy falls, and in so doing we can perhaps reduce platform interference potentials in radio bands.

Figure 2.39 is an example spreadsheet of a display frame header. It is composed of a finite set of symbols associated with blanking and sync functions. A typical feature of these

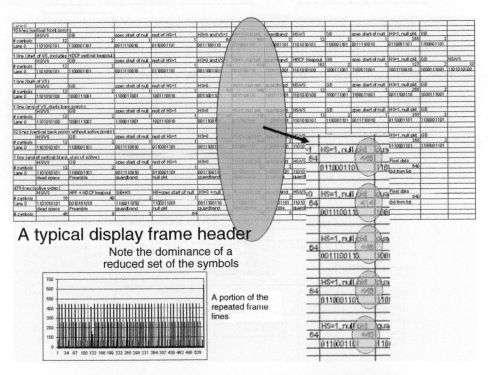

Figure 2.39: Symbol structure of a typical display frame.

header frames is that they are dominated by a few of the symbols of the set. These header frames might be composed of symbol sets comprising 20 or more individual symbols, but only a few of these symbols are dominant and are repeated most often. When we study the entries, we notice that some symbols are repeated a few times, some are repeated several hundred times, then there's another symbol sequence, and then a symbol is repeated several hundred times again. Stringing these symbol sequences out in time, one might think that creating the spectrum that goes along with such a complex symbol stream is a rather daunting prospect. However, one doesn't need to consider the effects of all of the symbols. Because some of the symbols in the set are only briefly on and then off for long periods of time, their associated spectra are much lower in amplitude and effect than are the symbols that are on longest. Because of this, we need to only consider the dominant symbols.

Figure 2.40 is an example of the display logic from notebook base to LCD display. Figure 2.41 is a schematic indicating the structure of each display page. The pixel data, the picture you see on your screen, is a subset of what is sent for each of the frames per

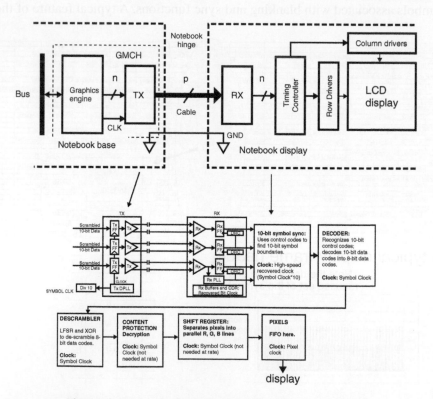

Figure 2.40: Logical structure of the display channel.

second. In order to ensure the stability of the picture, each page has a display frame format that surrounds the picture data.

So why do we care about all these display details? We care because the repeated symbol signal structure can in some cases look quite clock like. And the closer this structure comes to looking like a clock, the more the spectrum it produces looks like the spectrum a clock produces (and therefore increases the probability of interference in radio bands). Knowing how the symbol structure creates associated spectra provides us with yet another knob to turn to mitigate the impact of platform functions on wireless performance.

Figure 2.42 shows the relative distribution of energy between the symbols of the frame. So, while there are many symbols in the set, only a few symbols make up 99% of the spectral energy with one symbol composing 85%.

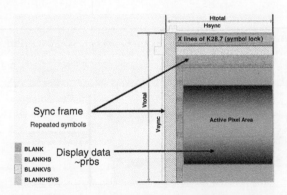

Figure 2.41: Schematic structure of each repeated display frame.

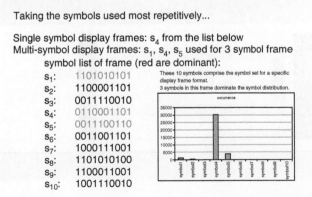

Figure 2.42: Energy distribution of the symbols in the display set.

Since only three of the symbols compose 99% of the spectral energy, we'll take a closer look at these symbols. We will construct a three-symbol sequence with just these symbols, and we'll use the relative frequencies of each symbol. Figure 2.43 shows the three symbols and their relative time durations.

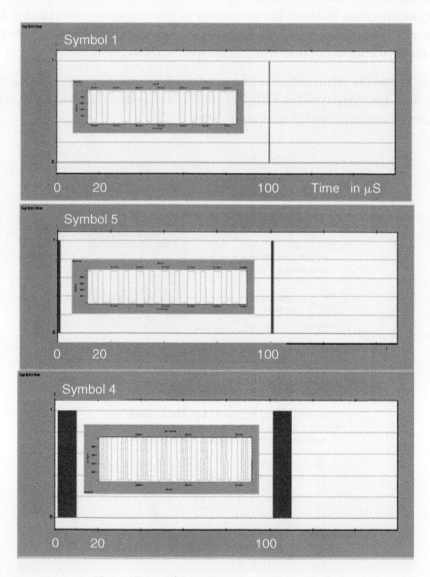

Figure 2.43: Time structure of the symbols.

The empty space between repetitions is where the display data would reside.

Figure 2.44 shows the combined sequence and the resulting spectrum. A noteworthy feature of this sequence, and indeed all similar sequences, is the appearance of spectral "holes," places in the spectrum where harmonics are absent. We should be able to take advantage of this spectral structure when considering platform design. We will consider this approach in greater detail later in this chapter. The point to be made now is that we can structure the display frame headers in such a way that we create these spectral "holes" at our wireless radio locations.

Figure 2.45 shows how the energy is distributed in time at a specific frequency. We take the signals and put them through a band-pass filter (Figure 2.46) and then show the

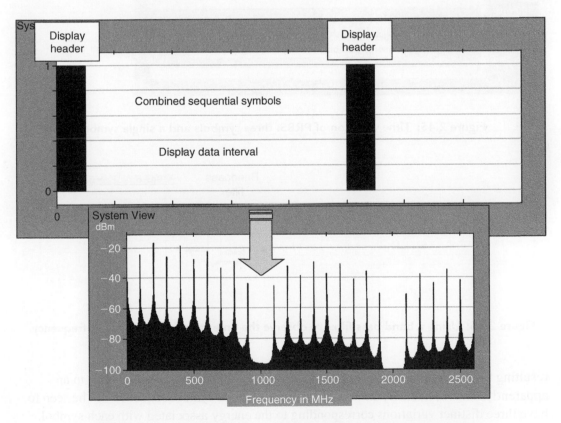

Figure 2.44: Combining the symbols and the resulting spectrum.

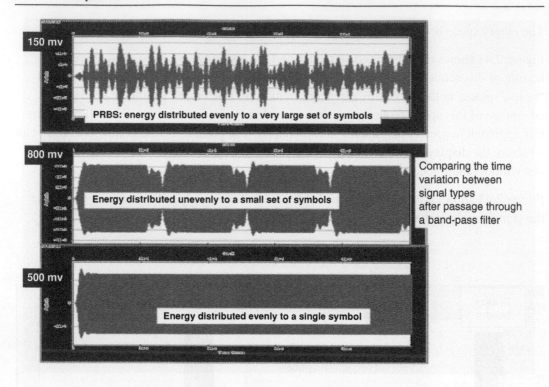

Figure 2.45: Time variation of PRBS: three symbols and a single symbol.

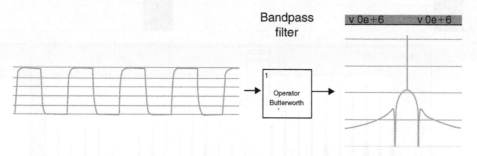

Figure 2.46: Using a band-pass filter to observe the time variation at a specific frequency.

resulting time variation. As might be expected for the PRBS, the energy varies in an apparently random manner; the distribution for the three symbol sequence can be seen to have three distinct variations corresponding to the energy associated with each symbol. The single symbol, of course, does not vary at all in time but is constant.

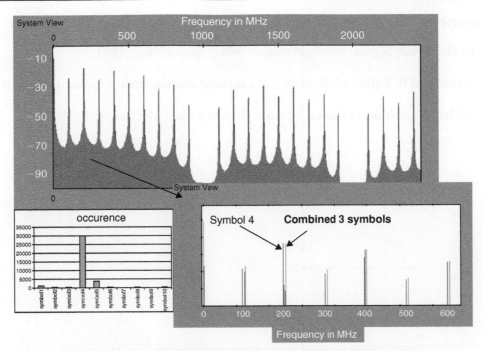

Comparing the 3 symbol sequence against the doimnant single symbol

Figure 2.47: Spectral comparison of the three-symbol sequence and the single symbol.

In Figure 2.47, the single symbol spectrum is nearly the same as the spectrum produced by the three symbols. This should be expected; the combined spectrum is dominated by the symbol that composes 85% of the total energy in the signal.

2.9 Summary

In this chapter, we have attempted to show the many different aspects of the structure of signals and how each element of this complex structure can be addressed to develop a strategy toward minimizing the interference impact of the signals that we use in our platforms. We have shown that edge rate asymmetry will produce even harmonics, where a symmetric signal does not. We have discussed how duty cycle can impact amplitude and variation in harmonic components of signals. We have also examined repetitive signals, non-repetitive signals, clocks that have been scrambled, and clocks that have been spread out in time in order to lower their associated harmonic amplitudes. Finally, we investigated the various symbols that are used to construct video digital displays.

References

[1] C.C. Goodyear, *Signals and Information,* Wiley-Interscience, 1971.

[2] F. Byron and R. Fuller, *Mathematics of Classical and Quantum Physics,* Dover, 1970.

[3] H.P. Hsu, *Signals and Systems,* Schaum Outline Series, McGraw-Hill, 1995.

CHAPTER 3

Analysis of Symbols

3.1 When People Finally Come to Understand Your Vision . . . It's Time for a New Vision

Digital display frames and their associated symbols present a problem to the EMI engineer. Their spectra fall somewhere between a purely repetitive symbol sequence such as a clock, and a purely random sequence of a finite set of symbols such as a pseudorandom bit stream (PRBS). The purpose of this section is to present a method whereby a selected set of display symbols can be ranked according to the expected EMI impact of their bit structure. The symbol set will be compared against the base clock from which the symbol is derived. The analysis will show that out of a given symbol set, a subset of the symbols should be used for highly repetitive sequences such as blanking to produce emissions that can be lower by up to 10 dB.

We will first look at single-ended symbols as seen at the input to a spectrum analyzer. We'll then discuss the time derivative spectra of single-ended symbols and then differential signals and their radiated emissions. Figures 3.1 to 3.8 show the spectra associated with various bit combinations of a 10 bit symbol. Figure 3.1 is the bit structure for a 50% duty cycle clock with an asymmetry between rise and fall times.

Note that the difference between even and odd harmonics is ~35 dB. Judging from this preliminary selection of spectra of symbols, we should expect quite wide variations across a symbol set.

Figure 3.6 shows the spectra produced by a display symbol, a 100 MHz clock (a single display symbol) and a PRBS at 1,000 MHz. The clock is assumed to be at 1/10 the data rate of the symbol and the PRBS, and all signals are assumed to be in phase. The amplitude equals 1 V_{pk-pk}, rise and fall time are equal (100 pS), and the clock is further assumed to be symmetric with a 50% duty cycle.

Figure 3.7 shows the time domain signals of the three signal types of Figure 3.6. Note that the clock has no even harmonics, and the symbol has both even and odd harmonics all at

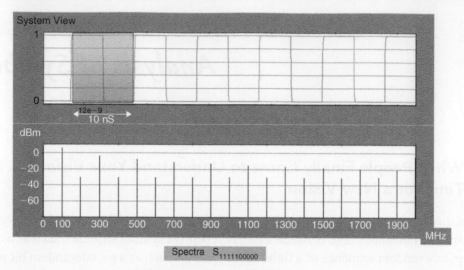

Rise = 200 pS, fall = 90 pS

The 1/10 data rate clock symbol

Figure 3.1: Bit structure for a 50% duty cycle clock. Note that even harmonics are present due to asymmetry between rising and falling edge rates.

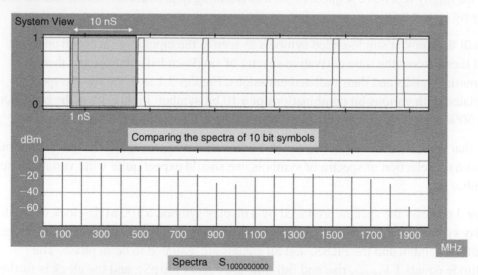

Our 10 bit symbol is made up of 10 1 bit pulses added to make a given symbol, therefore, all harmonics are present, even and odd. When we add them, the peak distribution among the harmonics changes.

Figure 3.2: Single bit spectrum.

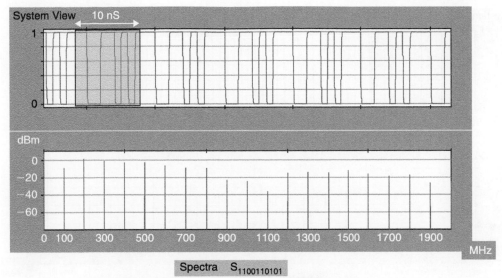

Spectra $S_{1100110101}$

Looking at the spectra derived from the 10 bit symbols.

Figure 3.3: Random symbol spectrum.

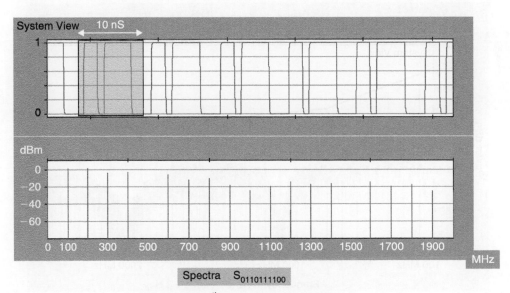

Spectra $S_{0110111100}$

Note that the 5th harmonic is now gone.

Figure 3.4: 10 bit symbol spectrum.

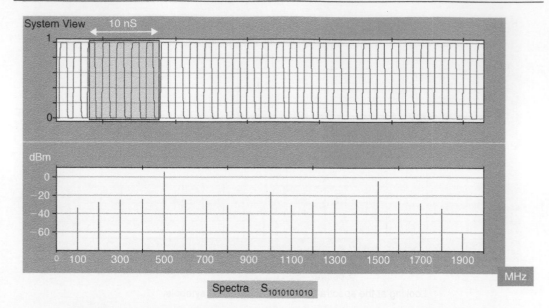

Figure 3.5: Filling in the bit structure to produce the highest clock rate.

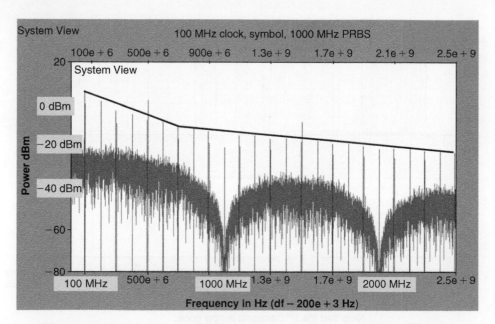

Figure 3.6: Comparison of spectra from clock, display symbol, and PRBS.

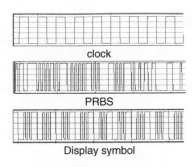

Figure 3.7: Clock, PRBS, and display symbol.

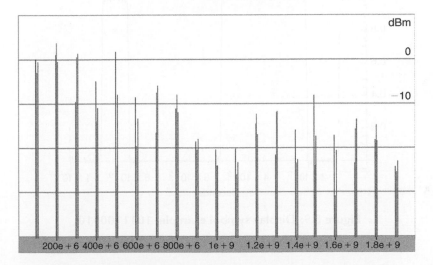

Figure 3.8: A comparison of the spectra for three symbols.

the fundamental clock spacing of 100 MHz. The clock fundamental has the highest peak of the spectrum set, the PRBS has the lowest peak set, and the symbol has some spectral components that fall above local clock spectral peaks.

The best approach to producing the lowest emissions profile is to scramble both the data stream and the clock, thereby producing spectra similar to the PRBS, which has the lowest set of spectral peaks. In some if not most application instances, this may not be practical or allowed. The question then is to determine which symbols of a given set will produce the lowest emissions.

Figure 3.8 shows a comparison of the spectra produced by three symbols with different bit patterns. Variations at particular harmonics can be as high as 25 dB. One symbol produces a maximum peak that is 4 dB higher than any other harmonic peak in the symbol set, and one symbol produces a harmonic set with peaks that are on average much lower than the other two symbols.

Figures 3.9 and 3.10 show typical time domain signals for a display symbol and a clock.

Any symbol can be decomposed into a set of its Fourier components. Figure 3.11 shows a symbol and its decomposition into three regions of time behavior. It is assumed that rise

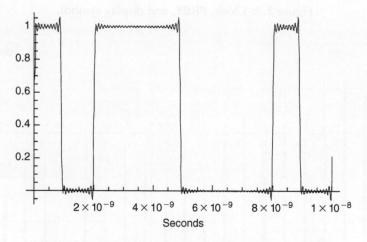

Figure 3.9: Display symbol example, 1011100010.

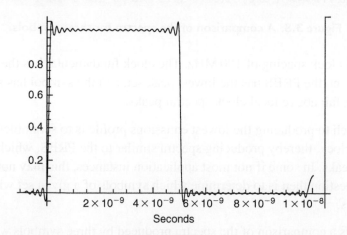

Figure 3.10: 50% duty cycle clock, 1111100000.

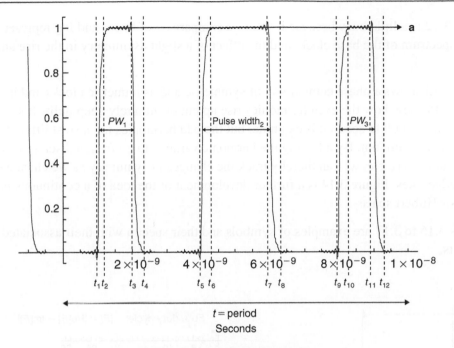

Figure 3.11: Display symbol.

time and fall time are always equal. We use this time structure to derive the set of Fourier components for each symbol of the display set.

The set of Fourier components can be viewed as a complex vector in C^n space, where n is the number of harmonics determining the dimension of the complex space.

We then take the inner product,

$$\left\| (v_1, v_2, v_3 \ldots v_n) \right\| = \left| \sqrt{(v_1, v_2, v_3 \ldots v_n) \cdot (v_1, v_2, v_3 \ldots v_n)} \right|$$

$$\text{where} \quad v_1 = (a_1 \pm jb_1),$$

$$v_2 = (a_2 \pm jb_2)$$

a and b are the harmonic Fourier components of the symbol.

The inner product is chosen because it is conceptually easy to appreciate, and it is an invariant of any given vector. By taking the real component of the inner product, the set of symbols can be ordered and ranked.

Figure 3.12 is a little busy because it shows a comparison of linear and log representations of the spectrum of the base clock, and the effect of a slight asymmetry in the rise and fall times.

Figure 3.13 shows a phase comparison of symmetric and asymmetric clocks and is intended to show how the even harmonics rise up out of infinitely deep nulls. It's somewhat difficult to see in this example, but the odd harmonics are also shifting, though by a very little amount. By allowing the harmonic variable to go from a discrete variable to a continuous variable, we can thereby track the changes in symmetry as we change either of the edge rates. Figure 3.14 is a further development of the idea of a continuous Fourier space, or Hilbert space.

Figures 3.15 to 3.17 are examples of symbols and their spectra with their associated inner products.

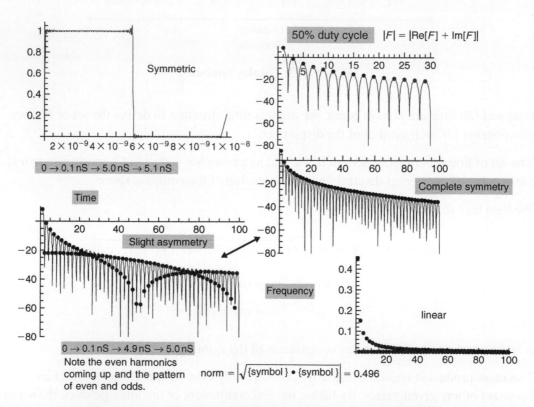

Figure 3.12: 50% duty cycle clock and its Fourier components.

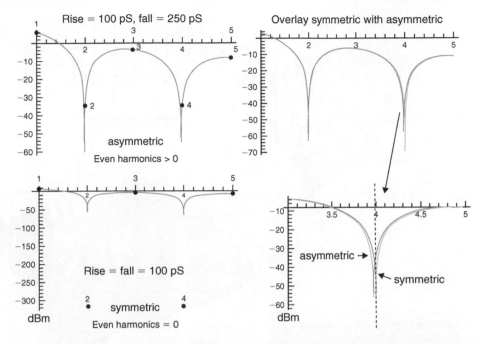

Figure 3.13: Phase shift of harmonics between symmetric and asymmetric clocks.

Table 3.1 lists a symbol set used for HDMI encoding. The symbols are composed of 2 and 3 event patterns, an event being defined as the presence of a "1" state (note that the symbols are DC-balanced). Table 3.2 shows the ordering of this symbol list using the inner product function. In this first ordering, comparing the single-ended spectrum as it would be seen at the input to a spectrum analyzer, the clock symbol is seen to have the highest inner product. A single bit symbol, representing a 10% duty cycle signal has the lowest.

We'll now show a set of measurements that compare single-ended spectra as generated by a pattern generator and input to a spectrum analyzer. Figure 3.18 shows the spectrum for the symmetric clock. The data comprises a set of 400 data points. However, most of the measured points are at or very near zero when measuring a clock signal. This is a problem that needs to be considered when collecting measurements. The larger the data set, the more accurate the measurement; however, more data points can lead to a sparse set of important data points. Figure 3.19 shows the result when we compare the clock symbol and the second symbol on the inner product list by subtracting one from another. We observe that the clock and the symbol are roughly equivalent across the harmonic range. If we take the larger data set and sort by amplitude, we can reduce the number of

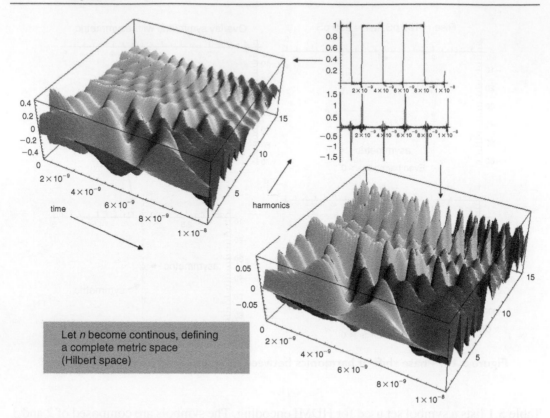

Figure 3.14: Analyzing the Fourier space of a symbol when the harmonic variable is allowed to become continuous.

comparative points. We can see the result in Figure 3.20. The areas under the positive and negative portions of the curves are roughly equal, and we can infer that the clock and the symbol will have similar spectral effects and therefore similar EMI effects.

Figure 3.21 shows a comparison of the highest inner product symbol, SymbolR, with a symbol further down the list, t0100111100, SymbolD. Performing the same data set reduction as above, we see that SymbolR clearly has the greater spectral impact. This is shown in Figure 3.22.

To this point, we have focused our discussion on developing the EMI/RFI impact of a single symbol taken from a finite symbol set. We can use the same approach to develop a similar analysis for a display symbol stream composed of multiple symbols. Figure 3.23 shows the relative occurrence of symbols within a real display sync frame set (the same symbol set from Chapter 2).

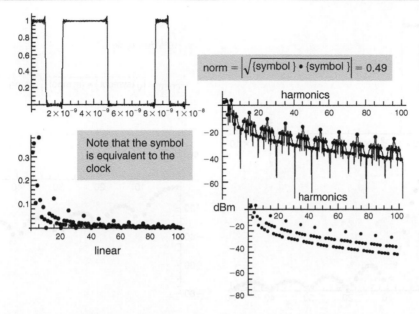

$$\text{norm} = \left| \sqrt{\{\text{symbol}\} \bullet \{\text{symbol}\}} \right| = 0.49$$

Note that the symbol is equivalent to the clock

Figure 3.15: Fourier components of a display symbol and its inner product.

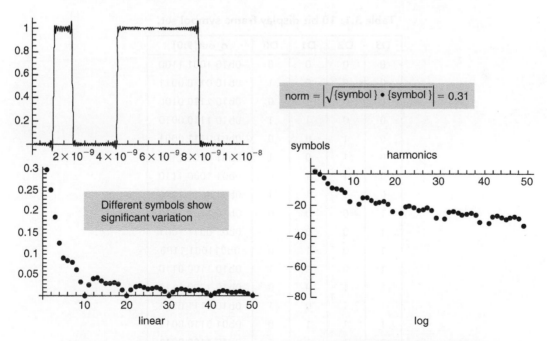

$$\text{norm} = \left| \sqrt{\{\text{symbol}\} \bullet \{\text{symbol}\}} \right| = 0.31$$

Different symbols show significant variation

Figure 3.16: Fourier components of a display symbol and its inner product.

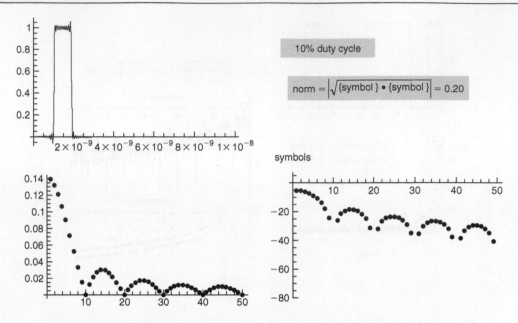

Figure 3.17: Fourier components of a display symbol and its inner product.

Table 3.1: 10 bit display frame symbol set.

D3	D2	D1	D0	q_out[9:0]
0	0	0	0	0b10 1001 1100
0	0	0	1	0b10 0110 0011
0	0	1	0	0b10 1110 0100
0	0	1	1	0b10 1110 0010
0	1	0	0	0b01 0111 0001
0	1	0	1	0b01 0001 1110
0	1	1	0	0b01 1000 1110
0	1	1	1	0b01 0011 1100
1	0	0	0	0b10 1100 1100
1	0	0	1	0b01 0011 1001
1	0	1	0	0b011001 1100
1	0	1	1	0b10 1100 0110
1	1	0	0	0b10 1000 1110
1	1	0	1	0b10 0111 0001
1	1	1	0	0b01 0110 0011
1	1	1	1	0b10 1100 0011

Table 3.2: Ordered symbol set according to Inner Product.

Symbol	Inner-Product Symbol
clock	0.496
t1011100010	0.49
t0110001110	0.418
t1001110001	0.418
t1011001100	0.383
t1001100011	0.376
t0110011100	0.376
t0100111001	0.347
t1011000110	0.347
t0101110001	0.307
t1010001110	0.307
80 percent	0.303
t1011100100	0.269
t0100111100	0.24
t1011000011	0.24
t100011110	0.237
t1010011100	0.225
t0101100011	0.225
single bit	0.2

As indicated, a small subset of the applied symbols makes up over 96% of the energy of the symbol set. With this in mind, a symbol algebra is shown below. It is limited to three symbols and the derivation of the inner product for that coupled symbol set. This method can be extended to a symbol stream of any length, in the limit approaching a true PRBS stream.

We designate the three symbols as S_1, S_2, S_3. Then,

$$S_1 = [u_1, u_2, u_3 \ldots u_n], \quad S_2 = [v_1, v_2, v_3 \ldots v_n], \quad S_3 = [w_1, w_2, w_3 \ldots w_n]$$

where we consider the Fourier component set as a row vector for each symbol. Then, following the previous development, $\left(S_1^T \, S_2 \, S_3^T\right) \left(S_1^T \, S_2 \, S_3^T\right)^T$, where S^T is the transpose.

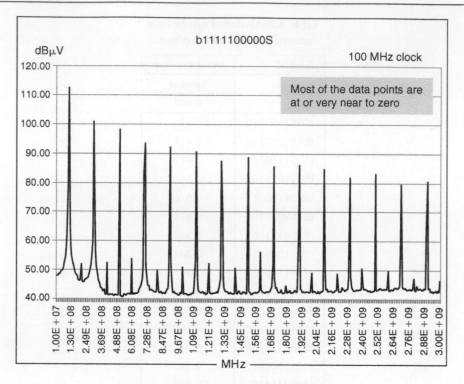

Figure 3.18: Measurement of the clock.

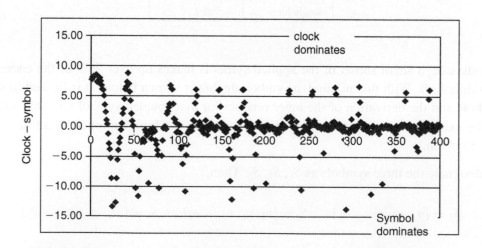

Figure 3.19: Comparing the clock and the highest inner product symbol.

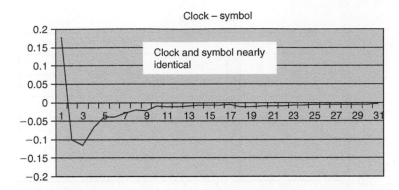

Figure 3.20: Reducing the data set and comparing effects.

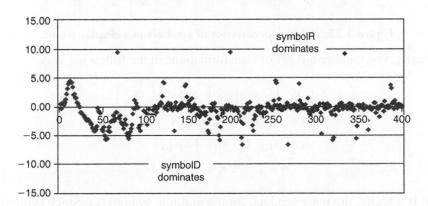

Figure 3.21: Comparing symbols for impact.

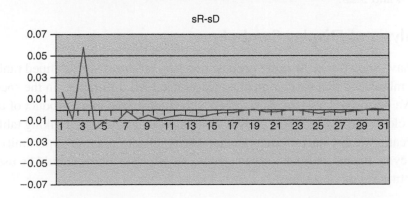

Figure 3.22: Reducing the data set and comparing effects.

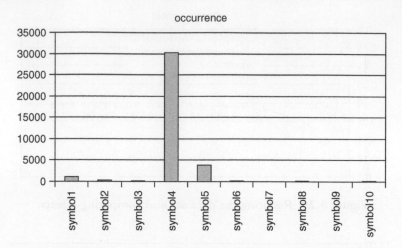

Figure 3.23: Relative occurrence of symbols in a display frame.

Geometrically, you can see this set of transformations in the following way.

$$\begin{bmatrix} S_1 \end{bmatrix}\begin{bmatrix} S_2 \end{bmatrix} \rightarrow \begin{bmatrix} S_1 S_2 \end{bmatrix}\begin{bmatrix} S_3 \end{bmatrix}$$
$$\underset{n \times 1}{} \underset{1 \times n}{} \qquad \underset{n \times n}{} \underset{n \times 1}{}$$

$$\rightarrow \begin{bmatrix} S_{123} \end{bmatrix}\begin{bmatrix} S_{123} \end{bmatrix} \rightarrow [\alpha]$$
$$\underset{n \times 1}{} \underset{1 \times n}{} \underset{1 \times 1}{}$$

The result is a scalar, the inner product, for the multiple symbol set. Since symbol algebra is not commutative, different temporal orders will produce different spectra, as shown in Figures 3.24 and 3.25.

3.2 Analysis of Display Symbols

So far we have shown that the inner product ordering produces well-defined ranking between symbols and that we can predict measurable EMI differences in the spectra of the symbols. We performed a set of measurements to compare the inner product of a symmetric clock and two symbols chosen from opposite ends of the ranking table. The results indicate that the inner product method allows the designer an opportunity early in the design cycle to choose from a given symbol set which symbols should be used for highly repetitive symbol streams.

This section will expand further on this approach to the ordering of spectral sets of digital display symbols. In particular, we will focus on developing the ranking and ordering of

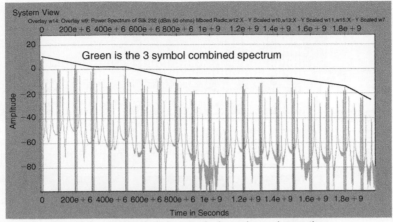

Combining the symbols in a particular order and repeating

Spectra $S_1 S_2 S_3$

Figure 3.24: Comparing temporal order impact on symbol spectra.

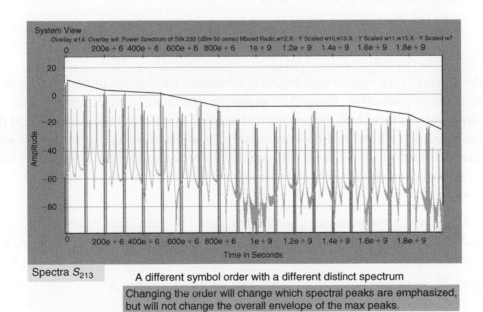

Spectra S_{213} A different symbol order with a different distinct spectrum

Changing the order will change which spectral peaks are emphasized, but will not change the overall envelope of the max peaks.

Figure 3.25: Comparing temporal order impact on symbol spectra.

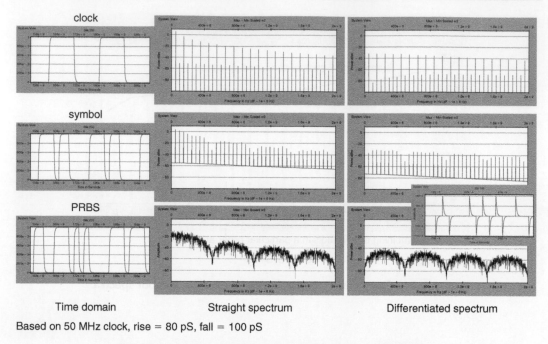

Time domain   Straight spectrum   Differentiated spectrum

Based on 50 MHz clock, rise = 80 pS, fall = 100 pS

The signal actually seen by the receiver will be differentiated

Figure 3.26: A comparison of the 3 signal types and their derivatives.

time derivatives of the symbol sets. Figure 3.26 shows a comparison of the three types of signals and their derivatives. Figure 3.27 shows the structure of a clock signal and its derivative. We will then show the ranking and ordering of differential signal pairs and their radiated emissions, where skew between signal pairs will assume an importance not seen in single-ended signals. Finally, we will examine the measurements of the impact of Gb/s display symbols in wireless bands and then show that the symbol set has an impact variation of ~15 dB.

Figure 3.27 shows a 100 MHz clock and its derivative. The amplitude equals 1 V_{pk-pk}, rise time and fall time are equal (100 pS), and the clock is further assumed to be symmetric with a 50% duty cycle.

The following equation describes the differentiated signal.

$$\partial_t F(t_r, t_f, pw, T, a, t) = \frac{\partial_t}{T} \left(\begin{array}{c} \int_0^{t_1} \frac{ate^{-\frac{j2\pi nt}{T}}}{t_1} dt + \int_{t_1}^{t_2} ae^{-\frac{j2\pi nt}{T}} dt \\ + \int_{t_2}^{t_3} \frac{a(t_3-t)e^{-\frac{j2\pi nt}{T}}}{t_3-t_2} dt \end{array} \right)$$

Recall that

$$\left\| (v_1, v_2, v_3 \ldots v_n) \right\| = \left| \sqrt{(v_1, v_2, v_3 \ldots v_n) \cdot (v_1, v_2, v_3 \ldots v_n)} \right|$$

$$\text{where} \quad v_1 = (a_1 \pm jb_1),$$

$$v_2 = (a_2 \pm jb_2)$$

a and b are the harmonic Fourier components of the symbol.

Figures 3.28a and 3.28b show both the real and imaginary components of the harmonic set.

Figures 3.29 to 3.34 show various symbols with rank and order spanning the range with the greatest EMI impact to those with the least EMI impact.

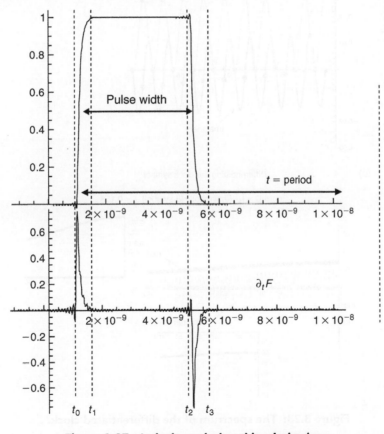

Figure 3.27: A clock symbol and its derivative.

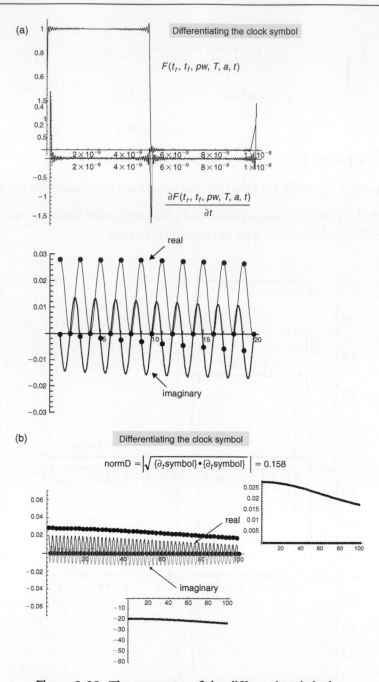

Figure 3.28: The spectrum of the differentiated clock.

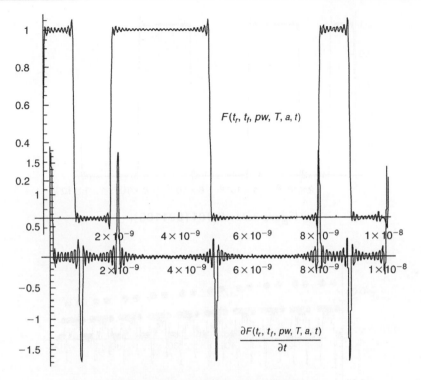

Figure 3.29: Symbol 1011100010.

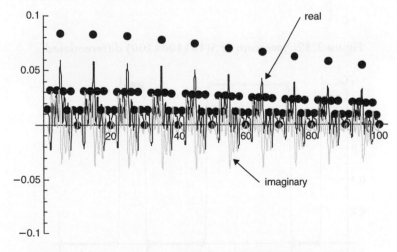

Figure 3.30: Spectrum of S(1011100010) differentiated.

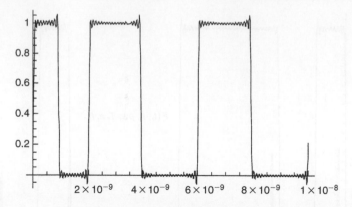

Figure 3.31: S(1011001100).

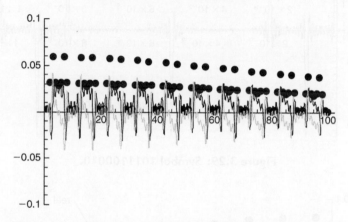

Figure 3.32: Spectrum of S(1011001100) differentiated.

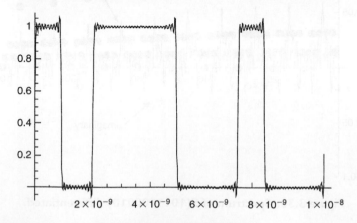

Figure 3.33: S(1011100100).

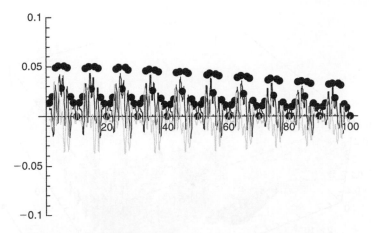

Figure 3.34: Spectrum of S(1011100100) differentiated.

The spectral structure of the symbols varies in a most interesting manner and deserves closer examination.

$$norm\,D = \left| \sqrt{\{\partial_t\,symbol\}\cdot\{\partial_t\,symbol\}} \right| = 0.273$$

$$norm\,D = \left| \sqrt{\{\partial_t\,symbol\}\cdot\{\partial_t\,symbol\}} \right| = 0.195$$

$$norm\,D = \left| \sqrt{\{\partial_t\,symbol\}\cdot\{\partial_t\,symbol\}} \right| = 0.037$$

Compare Figure 3.29 and Figure 3.33. They differ only in the placement of a single bit by 1 nano-second, yet their inner products are significantly different.

Figures 3.35 and 3.36 show the continuous distribution of harmonics for a symbol S(1011001100). Figure 3.35 is for the non-differentiated symbol, and Figure 3.36 is for the differentiated symbol. A surface has been generated by allowing harmonic number *n* to become continuous, defining a complete metric space (for easier visualization).

Table 3.3 is reproduced from before, and Table 3.4 is the result of ranking and ordering the inner product of the differentiated symbols.

As Table 3.4 indicates, the clock is no longer the dominant symbol, but instead is shown to have a 5 dB less EMI impact than does the highest inner product symbol. Figure 3.37 is a graphic ordering of Table 3.4 and clearly shows the range of the inner products between the symbols.

The question to be asked and addressed is: Given a symbol set and that certain symbols are chosen for repetitive syncing or blanking operations, is there a subset of the

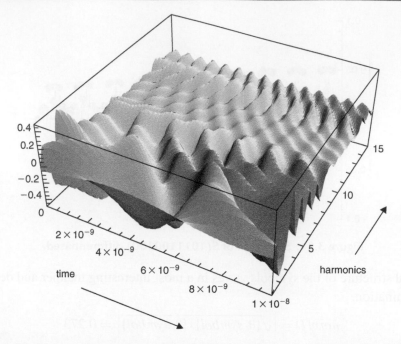

Figure 3.35: The continuous harmonic distribution for a non-differentiated symbol.

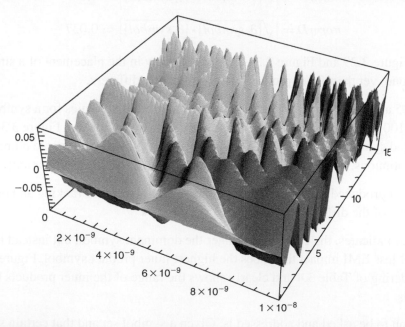

Figure 3.36: The continuous harmonic distribution for a differentiated symbol.

Table 3.3: 10 bit display frame symbol set.

D3	D2	D1	D0	q_out[9:0]
0	0	0	0	0b10 1001 1100
0	0	0	1	0b10 0110 0011
0	0	1	0	0b10 1110 0100
0	0	1	1	0b10 1110 0010
0	1	0	0	0b01 0111 0001
0	1	0	1	0b01 0001 1110
0	1	1	0	0b01 1000 1110
0	1	1	1	0b01 0011 1100
1	0	0	0	0b10 1100 1100
1	0	0	1	0b01 0011 1001
1	0	1	0	0b01 1001 1100
1	0	1	1	0b10 1100 0110
1	1	0	0	0b10 1000 1110
1	1	0	1	0b10 0111 0001
1	1	1	0	0b01 0110 0011
1	1	1	1	0b10 1100 0011

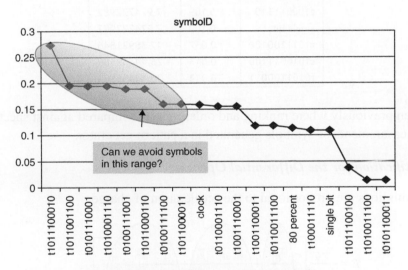

Figure 3.37: Distribution of the symbols.

symbols that should be chosen for these operations? The answer would appear to be a resounding yes!

Figure 3.38 is a comparison of the analytical results with a circuit simulation built to generate the symbols and then measure the resulting spectra. As Figure 3.38 shows, there is a quite good agreement between the two types of analysis. This further enforces the

**Table 3.4: Ordered symbol set according
to Inner Product.**

	symbolID	dB margin gain
t1011100010	0.273	0
t1011001100	0.195	2.922560714
t0101110001	0.194	2.967218342
t1010001110	0.194	2.967218342
t0100111001	0.189	3.194016857
t1011000110	0.189	3.194016857
t0100111100	0.159	4.695310454
t1011000011	0.159	4.695310454
clock	0.158	4.750111202
t0110001110	0.155	4.916618977
t1001110001	0.155	4.916618977
t1001100011	0.118	7.285612795
t0110011100	0.118	7.285612795
80 percent	0.113	7.661684071
t100011110	0.109	7.974722982
single bit	0.109	7.974722982
t1011100100	0.037	17.35921846
t1010011100	0.013	26.44438589
t0101100011	0.013	26.44438589

results given previously where ranking and ordering was compared against spectrum
analyzer measurements of symbols generated in a pattern generator.

3.2.1 Justification for the Differential Operator

Radiated emissions are proportional to the acceleration of charge.

$$E = \frac{-q}{4\pi\epsilon_0} \left[\frac{\vec{r}}{r^2} + \frac{r}{c}\frac{d}{dt}\frac{\vec{r}}{r^2} + \frac{1}{c^2}\frac{d^2}{dt^2}\vec{r} \right]$$

Coulomb
term

Steady
current
term

Radiated
term

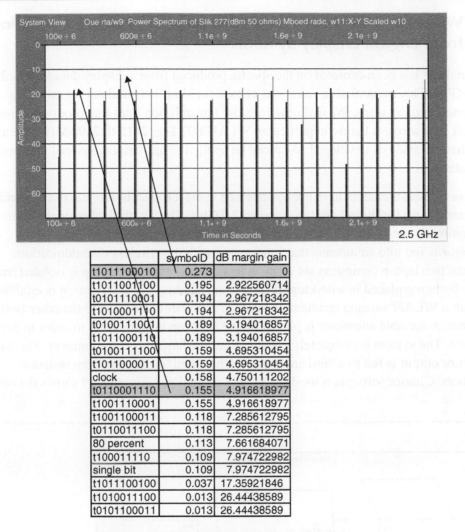

	symbolID	dB margin gain
t1011100010	0.273	0
t1011001100	0.195	2.922560714
t0101110001	0.194	2.967218342
t1010001110	0.194	2.967218342
t0100111001	0.189	3.194016857
t1011000110	0.189	3.194016857
t0100111100	0.159	4.695310454
t1011000011	0.159	4.695310454
clock	0.158	4.750111202
t0110001110	0.155	4.916618977
t1001110001	0.155	4.916618977
t1001100011	0.118	7.285612795
t0110011100	0.118	7.285612795
80 percent	0.113	7.661684071
t100011110	0.109	7.974722982
single bit	0.109	7.974722982
t1011100100	0.037	17.35921846
t1010011100	0.013	26.44438589
t0101100011	0.013	26.44438589

Figure 3.38: Comparing analysis to simulation.

And,

$$V = IR = R\frac{dq}{dt}$$

$$\frac{dV}{dt} = R\frac{d^2q}{dt^2}.$$

Therefore, the radiated emissions are associated with the symbol edges.

3.3 Wireless Performance in the Presence of Radiated Emissions from Digital Display Symbols

Our analysis has been centered on the spectra produced from a display data set running at 1,000 Gb/s. This transmission value was chosen for simplicity and not for any deep implementation reasons. We will now consider transmission rates that produce emissions that fall into wireless bands, in particular, WLAN 802.11g, 2412 to 2490 MHz. We assert here that the analysis developed so far can be applied to any display symbol set running at any data rate.

A series of measurements have been performed using a transmission rate that is coming into general use for digital display systems: 2.7 Gb/s. The set of display symbols are programmed, one at a time, into a 3.3 GHz pattern generator, and the output of the generator is fed into an antenna that can interact with the wireless communications between two laptop computers set up in an ad hoc manner. Each laptop is isolated from the others by being placed in a desktop RF isolation chamber. Communication is established through a WLAN antenna mounted inside each chamber and cabled to the other isolation chamber. A variable attenuator is placed in line between the chambers in order to simulate distance. The system is completely isolated from the laboratory environment. The pattern generator output is fed to a third antenna, which is then fed to one of the isolation chambers. Chariot software is used to measure throughput. Figure 3.39 shows the results

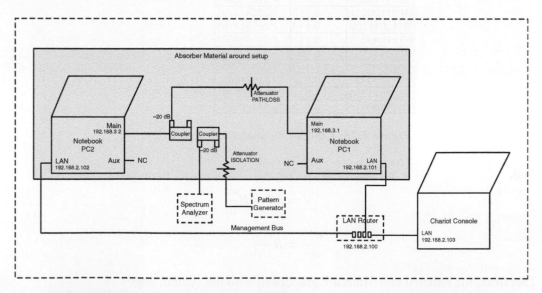

Figure 3.39: Test set up to measure interference potentials in wireless channel.

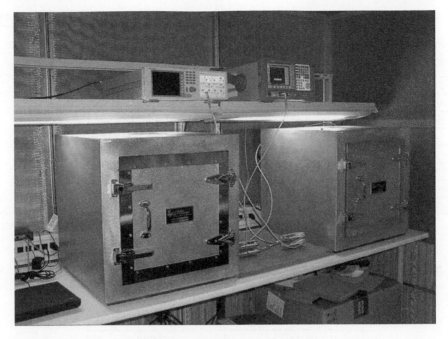

Figure 3.40: The test set up.

of the experiment. For these measurements, the pattern generator output is single-ended. Subsequent analysis will consider differential symbols.

Figure 3.40 is a photo of the experimental setup. Figure 3.41 shows a plot of the results.

The curve to the far right is throughput in the absence of injected noise. The fall off shows several "ledges" where the fall off abates and is constant and then starts falling off again. These ledges are the result of rate algorithm adjustments. The rate is adjusting even in the absence of noise due to the radio operating at the extreme of its sensitivity. Hence, as attenuation continues to be dialed in, the rate algorithm will adjust downward attempting to maintain the link (albeit at reduced rates), until it can no longer maintain the link.

The other curves indicate the successful transmission rate fall off in the presence of the various symbols.

Table 3.5 shows a comparison between the measured results and the ranking derived from using the inner product of the complex harmonic set.

The analysis indicates a single symbol having minimal impact and clearly separated from the rest of the symbol set. The measured results show the same. The analysis indicates a spread of 14 dB between the least and greatest symbol, and the measured results show

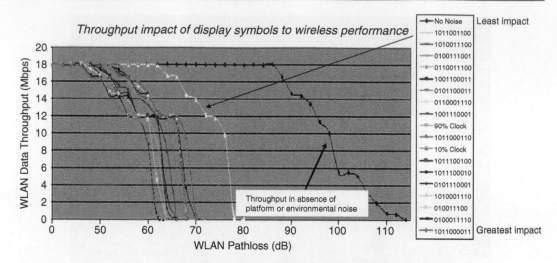

Figure 3.41: Comparison of test results.

Table 3.5: A clock symbol and its derivative.

Measured ranking	Analytical ranking
1011001100	0.009
1010011100	0.023
0100111001	0.023
0110011100	0.023
1001100011	0.023
0101100011	0.023
0110001110	0.023
1001110001	0.023
1011000110	0.023
1011100100	0.035
1011100010	0.036
0101110001	0.036
1010001110	0.0366
0100111100	0.048
0100011110	0.048
1011000011	0.048

the same. So, it is clear that the method of inner product ranking has merit and can be correlated with measurement of real performance impact. We will now consider differential pairs of transmitted symbols.

3.4 Developing the Analysis of Differential Symbols

We will begin the analysis with the time domain description of the 50% duty cycle clock. Figure 3.42 shows a differential clock with no skew. Figure 3.43 shows the differential clock with 200 pS of skew added to the D-side. For illustrative reasons, we've shown 200 pS of skew. Anything less is somewhat difficult to see when viewing the entire clock signal. We will show that skew values of 10 pS can cause measurable levels of interference.

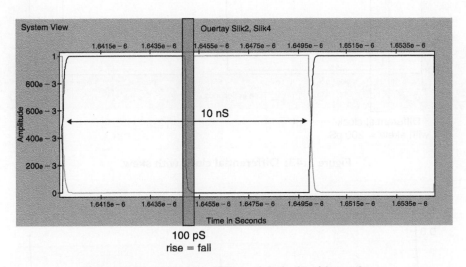

Figure 3.42: 100 MHz differential clock with no skew.

Figure 3.44 shows the time derivative of the two sides of the differential symbol added together (with 10 pS of skew).

Even with as little as 10 pS of skew, there remains a significant signal left to contribute to radiation. Figure 3.45 shows the resulting spectrum.

Figure 3.46 shows the effect of 100 pS of skew between D+ and D−. Figure 3.47 shows the resulting spectrum.

Simply put, the greater the skew the higher the emissions.

Following the symbol procedure previously described, a ranking is now derived for the differential symbols in order to predict their EMI impact and compare against single-ended predictions. This is shown in Table 3.6.

This comparison is shown graphically in Figure 3.48.

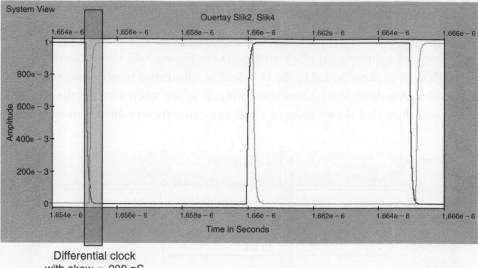

Differential clock
with skew = 200 pS

Figure 3.43: Differential clock with skew.

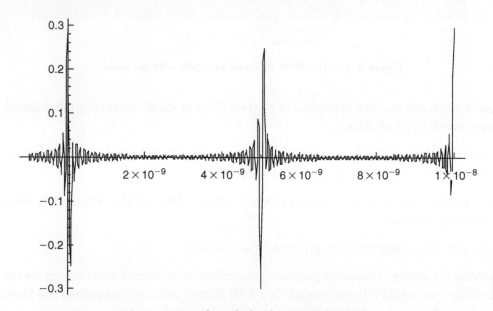

Figure 3.44: Time derivative of the symbol.

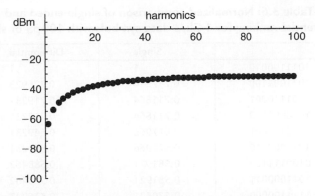

Figure 3.45: Spectrum for time-differentiated differential symbol.

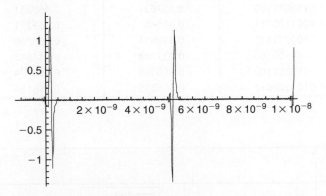

Figure 3.46: Differential signal with 100 pS of skew.

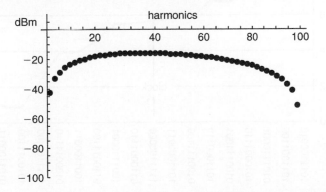

Figure 3.47: Spectrum for differential signal with 100 pS of skew.

Table 3.6: Normalized comparison of single-ended and differential ranking for 1 Gb/s data symbols with 100 pS of skew.

	Single	Differential
1011100010	1	1
1011001100	0.717514	0.726923
0101110001	0.711864	0.719231
1010001110	0.711864	0.719231
0100111001	0.677966	0.669231
1011000110	0.677966	0.669231
0100111100	0.581921	0.588462
1011000011	0.581921	0.588462
1111100000	0.570621	0.576923
0110001110	0.553672	0.546154
1001110001	0.553672	0.546154
0110011100	0.457627	0.469231
1001100011	0.39548	0.469231
0100011110	0.389831	0.380769
1011100100	0.237288	0.234615
1010011100	0.090395	0.096154
0101100011	0.090395	0.096154

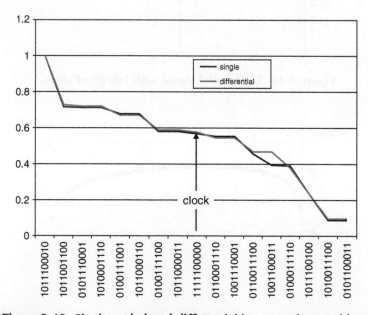

Figure 3.48: Single-ended and differential inner product ranking.

Table 3.7: Non-normalized comparison of single-ended and differential signal with 100 pS of skew.

	IPS(n,1,50)	IPD(n,1,50)100 pS
1011100010	0.177	0.26
1011001100	0.127	0.189
0101110001	0.126	0.187
1010001110	0.126	0.187
0100111001	0.12	0.174
1011000110	0.12	0.174
0100111100	0.103	0.153
1011000011	0.103	0.153
1111100000	0.101	0.15
0110001110	0.098	0.142
1001110001	0.098	0.142
0110011100	0.081	0.122
1001100011	0.07	0.122
0100011110	0.069	0.099
1011100100	0.042	0.061
1010011100	0.016	0.025
0101100011	0.016	0.025

Figure 3.48 indicates that the ranking order remains the same for each symbol whether single-ended or differential. Table 3.7 shows the non-normalized comparison.

As Table 3.7 shows, the differential symbols, while having the same rank order, produce a higher impact. Figure 3.49 shows the effect of varying the skew in the differential symbol from 10 pS to 200 pS.

The maximum impact of skew occurs when the skew is of the same magnitude as the rise and fall times of the symbol edges. Figure 3.50 shows the measured impact on wireless throughput due to differential symbols with skew. Measured and analytic results show good correlation. Especially note that 20 pS of skew will produce significant wireless throughput degradation.

We set out to show that a set of complex display symbols could be ordered through the use of the inner product of their Fourier set representation. The inner product ordering was shown to produce well-defined ranking between symbols and to predict measurable EMI differences in the spectra of the symbols.

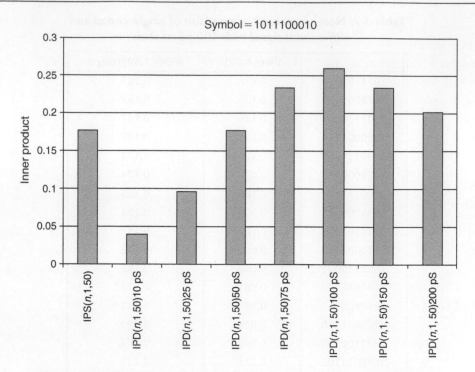

Figure 3.49: The effect of varying the skew in the differential symbol.

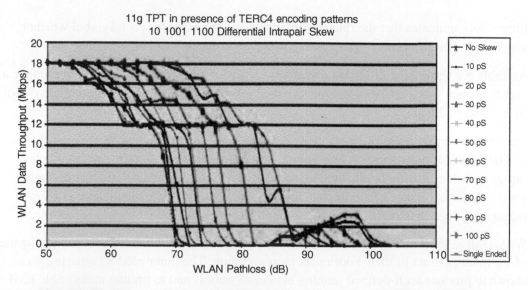

Figure 3.50: Measured impact on wireless throughput due to differential skew.

We then expanded on this approach to the ordering of spectral sets of digital display symbols by focusing on developing the ranking and ordering of time derivatives of the symbol sets. It was shown that the order and EMI impact changed considerably under the differentiation operator. The clock symbol was no longer the dominant symbol but instead fell to somewhere in the middle of the considered symbol set. The implication is that one cannot assume that a clock symbol is going to dominate in terms of EMI impact when the clock is part of an extended complex symbol set.

We have demonstrated that display symbols can have varying impact on wireless performance, and that the basic idea of ranking the EMI/RFI impact of the symbol set has merit when used in conjunction with the design of a display protocol.

Further, we have shown that when skew is present, differential symbols will have a greater impact on EMI/RFI than for a single-ended symbol if the skew is between 1/2 and 3/2 of the symbol edge rate.

3.5 The Impact of Pulse Width and Edge Rate Jitter on the Expected EMI/RFI

Returning to signal asymmetries, we will investigate jitter in the edge rates and in the duty cycle or pulse width. Jitter introduces variation in the resulting spectrum, in some cases creating harmonics where theory says they shouldn't be. We observed this when we introduced edge jitter into a pure PRBS signal and spikes arose at the data rate harmonics where there should have been quiet nulls.

Additionally, the entire idea of data rate nulls is misleading. Differential signaling has been introduced for signal integrity purposes and for reducing interference. As we've seen, though, skew can enhance radiated emissions from differential signals. In differential signal nulls, we see conversion of energy from differential mode to common mode. Figure 3.51 is a comparison of spectra between a signal without jitter and the same signal with jitter.

Note the complete absence of even harmonics in the signal without jitter and the appearance of even harmonics as soon as the edge rates become unequal. This is true for both the straight and differentiated signals.

Figure 3.52 takes a closer look at the mechanism of jitter in relation to phase variation.

In Figure 3.52, we've allowed the function describing the harmonic set to become continuous to better illustrate the behavior as the signal diverges from strict symmetry.

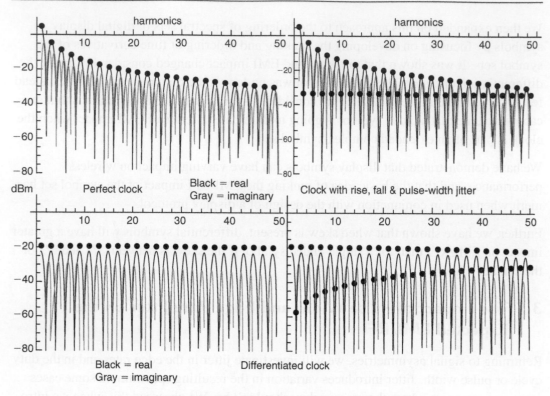

Figure 3.51: Differential signals with edge rate jitter.

Note that in the symmetric condition only odd harmonics exist; the even harmonics exist at the bottom of an infinitely deep null. When we introduce jitter to the signal (whether the jitter is in edge rate or pulse width), we create an asymmetry in the signal structure that effectively brings the even harmonics up and out of the null well and gives them a finite amplitude. This occurs whether we are considering the signal or the differentiated signal. In Figure 3.53, we look more closely at a particular harmonic to observe the difference between the perfect clock and the jittered clock. One way of looking at the whole thing is to think of the jitter as introducing a phase distortion into the signal. This distortion then shifts the signal relative to the perfect signal, which lands the even harmonics higher on the amplitude shoulder.

Figures 3.54 to 3.58 show the impact on the RFI potential of clocks and symbols when jitter is present in the signal. For this analysis, we created a Monte Carlo sort of set of runs and varied the rise and fall times of the signal edges and the pulse widths. We then

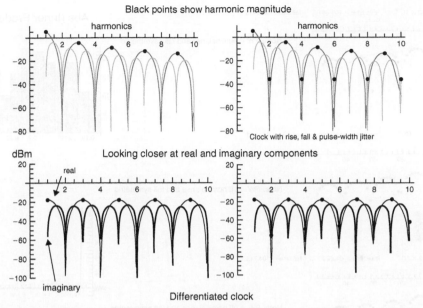

Figure 3.52: Closer examination of the real and imaginary components of a symmetric versus non-symmetric signal.

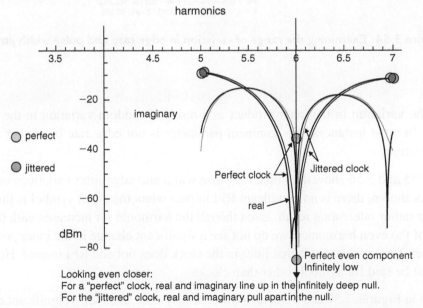

Figure 3.53: An asymmetric condition raises the even harmonic component out of an infinitely deep null.

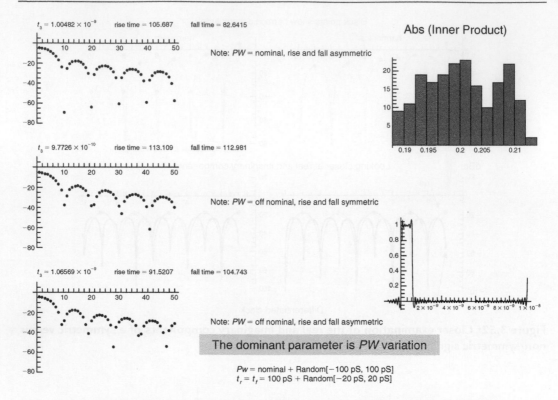

Figure 3.54: Examining the range of variation in edge rate and pulse width jitter.

showed the variation in the inner product as a result of random variation in the signal structure. In most instances, the dominant parameter is not edge rate but rather the pulse width variation.

Figures 3.55 and 3.56 show the impact of pulse width and edge jitter variations on the RFI impact. As shown, there is no significant RFI impact when the clock symbol is jittered, which is a rather interesting result. Even though the harmonic set increases with the addition of the even harmonics, we do not see a significant change in the inner product as a result. Therefore, we conclude that jitter in the clock does not add RFI impact. However, this cannot be said for symbols other than clocks.

As shown in Figures 3.57 and 3.58, an asymmetric symbol does see a significant change in its RFI impact.

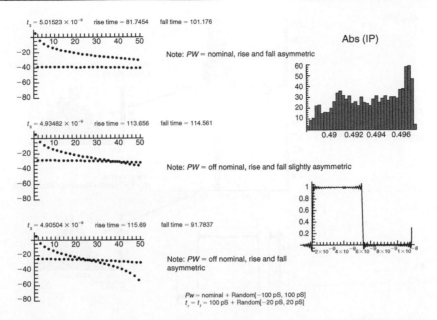

Figure 3.55: Examining the range of variation in edge rate and pulse width jitter in a symmetric signal.

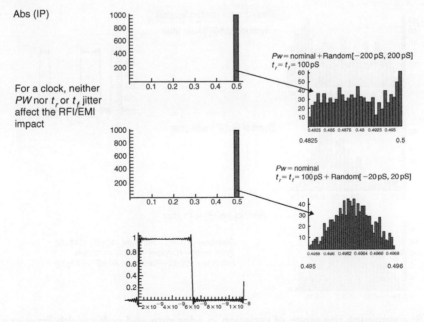

Figure 3.56: Examining the range of variation in edge rate and pulse width jitter in a symmetric signal.

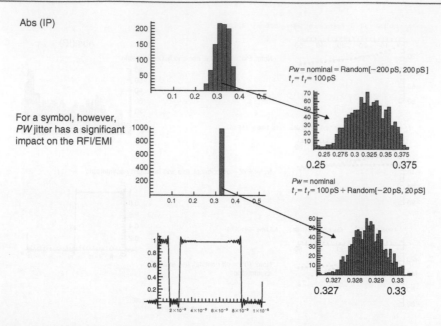

Abs (IP)

For a symbol, however,
PW jitter has a significant
impact on the RFI/EMI

Pw = nominal = Random[−200 pS, 200 pS]
$t_r = t_f = 100$ pS

Pw = nominal
$t_r = t_f = 100$ pS + Random[−20 pS, 20 pS]

Figure 3.57: Examining the range of variation in edge rate and pulse width jitter in a symbol other than a clock.

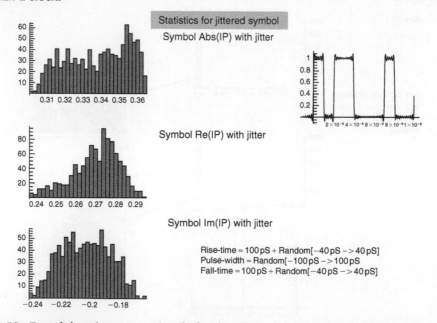

Statistics for jittered symbol
Symbol Abs(IP) with jitter

Symbol Re(IP) with jitter

Symbol Im(IP) with jitter

Rise-time = 100 pS + Random[−40 pS −> 40 pS]
Pulse-width = Random[−100 pS −> 100 pS]
Fall-time = 100 pS + Random[−40 pS −> 40 pS]

Figure 3.58: Examining the range of variation in edge rate and pulse width jitter in a symbol other than a clock (real and imaginary variation).

3.6 Summary

The point of this chapter has been to expand on the impact of signal structure in the platform and the variation in RFI as a result of that variation. We have attempted to show and suggest to the reader the many possible approaches to interference mitigation that may be pursued through application of signal structure techniques. We have also attempted to indicate those aspects of signal structure that may be ignored or may be a waste of time if it is solely for RFI improvement. In this bucket, we would include worrying about edge rate jitter when it is pulse width jitter that contributes the larger variance in RFI impact.

References

[1] C.C. Goodyear, *Signals and Information*, Wiley-Interscience, 1971.

[2] F. Byron and R. Fuller, *Mathematics of Classical and Quantum Physics*, Dover, 1970.

[3] H.P. Hsu, *Signals and Systems*, Schaum Outline Series, McGraw-Hill, 1995.

CHAPTER 4

Measurement Methods

4.1 One Measurement Is Worth a Thousand Opinions

Throughout this book, we have interspersed a number of measurement examples to help the reader to develop the engineering intuition. In this chapter, we will bring these various methods together in a single area of reference. At the same time, we'll develop methods of analysis and models that are meant to work hand in hand with the measurement methods.

The near-field scanner shown in Figure 4.1 was developed specifically to measure the magnetic and electric fields at the surface of integrated circuit (IC) packages. Because the system uses optical-quality stepper motor, it can be used to scan over IC dies as well, with a resulting spatial resolution of 50 microns. This method for close measurement of silicon and packages allows for developing a better understanding of radiation mechanisms in micro structures such as power delivery networks, clock distributions, and bonding wires. This understanding in turn leads to design guidelines at the package and silicon levels (Fig 4.1). Through better theory and analysis of the surface fields, we are led to better design guidelines that impact signal integrity as well as EMI and RFI.

The system shown in Figure 4.2 is a scanner developed at Intel. There are other systems available on the market from various vendors with varying levels of functionality and resolution.

In addition to scanning, there are different approaches to how the measured data can be used. We will describe several methods later in the chapter.

Figure 4.3 is a simple schematic of the near-field scanning system. System requirements are a typical PC platform for control, motor controllers and feedback, a spectrum analyzer, and a digital signal oscilloscope. While not necessary, a shield room is suggested. For quite some time, we did not require a shield room, but with the advent of ubiquitous wireless, we found that the lab environment had become quite noisy, especially if we wanted to scan for IC emissions in wireless bands.

Figure 4.1: Near-field scanning of PCBs, packages, and silicon.

The near-field scanner has been used for a number of years in investigation of PC system clock devices. The system clock is a significant source of system emissions. By understanding emission mechanisms in these devices and thereby developing mitigation schemes for the silicon and associated packages, we can reduce a large set of those signals that contribute to both EMI and RFI system problems. In many cases, by addressing silicon changes, we can produce significant EMI reduction at essentially zero cost.

The figures that follow illustrate some of the scanning results. Figure 4.4 is a set of electric field measurements at the surface of a system device. Most experienced workers in EMI/EMC know that a trapezoidal-type signal has two significant amplitude roll offs of 20 dB and 40 dB. These break points are shown in Figure 4.4. As we can see, the measurements indicate that the device conforms to what is usually expected from the

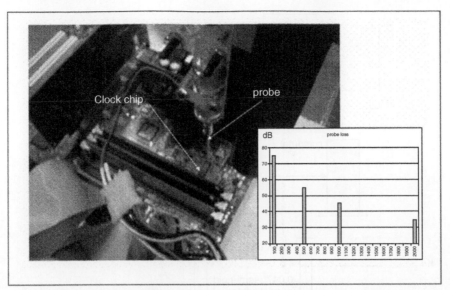

Figure 4.2: Near-field scanning of a system clock device and the probe factors associated with an electric field probe.

measured spectrum of devices with a symmetric trapezoidal signal excitation. We will later see that the story changes as device edge rates speed up to less than a nanosecond, which these days occurs in almost every device. In addition, hanging capacitors on the device outputs will probably do nothing to reduce the emissions if the emissions are emanating from the device package and lead frame. We'll discuss that later.

The figures that follow show scans over an assortment of system devices, such as the hard disk, the clock, the I/O control hub (ICH), the graphic and memory control hub (GMCH), and memory devices. As we can see, there is quite a spread in emission amplitudes and in field patterns. As we develop an understanding of these fields, we will build models that enable us to explore the radiation mechanisms of these devices and the methods for lowering the emissions of complex sets of radiation sources.

The Intel near-field scanner (INFS) is capable of both horizontal and vertical scans.

Figure 4.5 shows an example of what can be done with the INFS—measuring the electric field throughout a dielectric and across the boundary between two dielectrics. In this case, we placed a radiation source in a dielectric fluid and ran the INFS in the vertical measurement mode. Note the boundary effect where the source distribution is blurred spatially. This is an important result that we will discuss again.

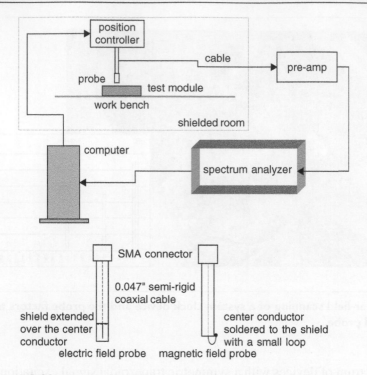

Figure 4.3: Near-field system schematic and associated electric and magnetic field probes.

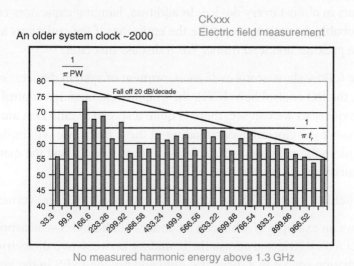

Figure 4.4: Near-field measurements of the electric field over a PC system clock device, circa 2000.

Propagation through a dielectric: $e_r = 2$

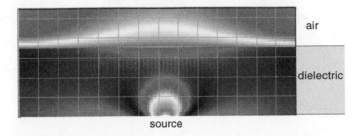

Figure 4.5: Measuring the electric field in a dielectric and across a boundary.

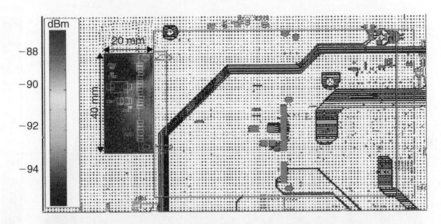

Figure 4.6: Horizontal near-field scan over a system hard disk.

As far as we know, this is the only measurement of this type.

In Figure 4.6, a horizontal near-field scan has been superimposed on the artwork for a notebook system. We can see that only a subset of the hard disk I/O pins is hot with emissions. If we can show a correlation between system level EMI/RFI and the frequencies measured at these pins, we can then spend our limited time examining these pins for required mitigation efforts.

When we look at the vertical plot of the same pins in Figure 4.7, we can see that they appear somewhat isotropic, showing no preferred direction. This indicates no gain in any direction.

Figure 4.7: Vertical near-field scan over a system hard disk.

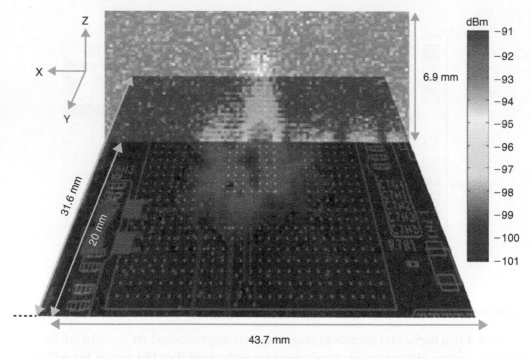

Figure 4.8: Combined vertical and horizontal scans over a system I/O control hub IC.

In Figure 4.8, we show the combined vertical and horizontal scans for an ICH chip. In this instance, we can see that the vertical emissions have a certain degree of directivity. Early in a design cycle, this information could give a platform designer an advantage in the placement of probable interference sources and interference victims. It can also lead to better system understanding of coupling mechanisms between different functional areas of the platform.

The near-field scanner can make good measurements in the low MHz to 10 GHz. Figure 4.9 shows a set of measurements performed in wireless bands over various types of silicon. Note the different types of field patterns. The first two are rather simple in structure; the lower two show a very complex structure. The question we will be asking is whether or not we can construct models that reproduce these different types of field patterns.

The added power of near-field scanning is the ability to make the scans transparent and combine them with system artwork. An example is shown in Figure 4.10, which is a scan over a clock device and its associated PCB circuitry. In Figure 4.11, we show a vertical scan of the same device. As can be seen in this plot, the vertical emissions are showing a significant directivity.

Okay, so what? What does the collection of near-field scanning do for us other than produce sets of quite pretty pictures? They are sometimes stunningly pretty pictures, but

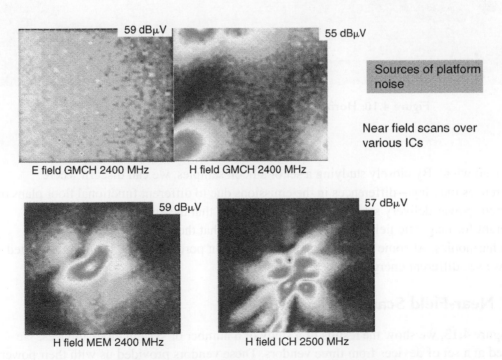

Figure 4.9: Near-field patterns over various IC types.

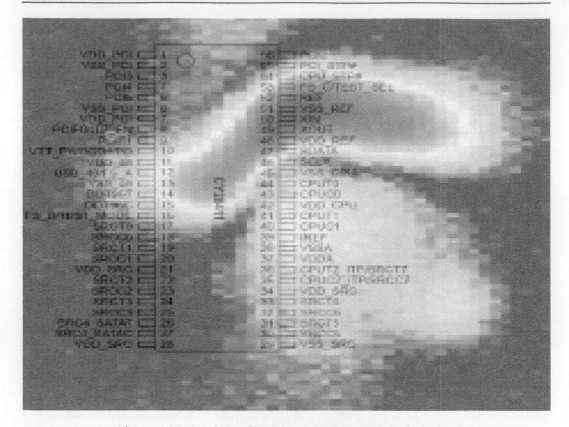

Figure 4.10: Horizontal scan over a PC system clock device.

again, so what? By closely studying a set of silicon samples, we can discern where differences may lay—differences in the emissions due to different functional floor plans or different power delivery schemes. Figure 4.12 shows that the energy distributions are different for magnetic fields and for electric fields—that they are different for odd and even harmonics. At some frequencies, we see different portions of the lead frame excited and we see different energy storage mechanisms.

4.2 Near-Field Scans of Clock ICs

In Figure 4.13, we show the results of a study of a number of system clock devices. We looked at a set of devices from three vendors. These vendors provided us with their power layer metal artwork. Each device met functional requirements as regards to power

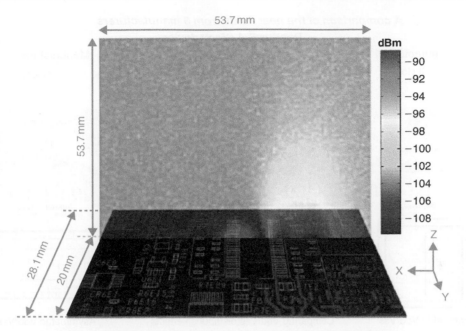

Figure 4.11: Vertical scan over a PC system clock device.

Near Field Scans of Clock IC

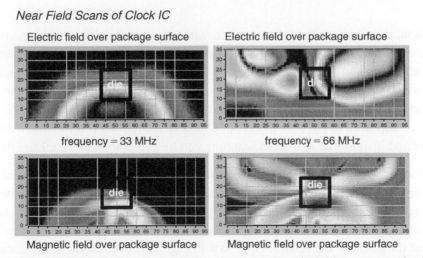

Figure 4.12: Varying field distributions for electric and magnetic field at even and odd harmonics. Fundamental clock frequency = 33.3 MHz.

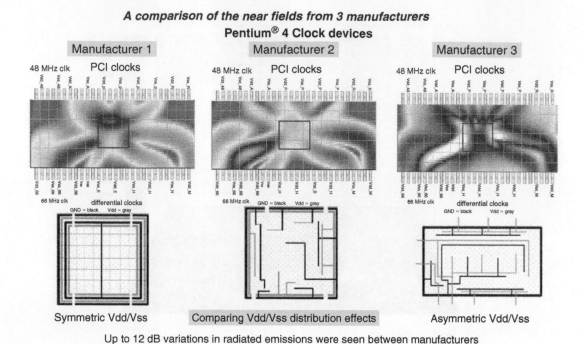

Figure 4.13: Comparison of the near-field emissions of three vendors for system clock devices.

consumption, signal integrity, and so on. Yet, when measured in PC platforms, we saw a difference of 12 dB in the measured radiated emissions in a 3 m chamber. Near-field scanning allowed us to determine that the power delivery network (PDN) for each device was the proximate cause of the EMI/EMC discrepancies. For the best device, the power delivery artwork indicated that designers paid attention to symmetry in the PDN, while the worst device appeared to have been laid out with an auto-routing routine. It's with this type of analysis and insight that near-field scanning can benefit silicon and package designers.

The scans we have shown so far are somewhat older, dating from an older generation of system clock devices. Figures 4.14 to 4.16 show more recent scans of a new generation of system clocks. For these devices, the vendor provided us with de-encapsulated parts that allowed us to get closer to the silicon itself and to also eliminate the blurring due to the presence of the package dielectric. As we saw in Figure 4.5, dielectric blurring can be a significant factor in reducing imaging capability. By being able to scan the silicon fields without the blurring effect of the dielectric, we can attain very good spatial resolution and

Device Under Test – CK505 sans package

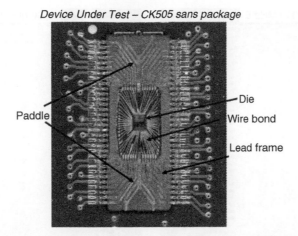

Figure 4.14: Scanning a clock device without package dielectric effects.

700 MHz Field Distribution

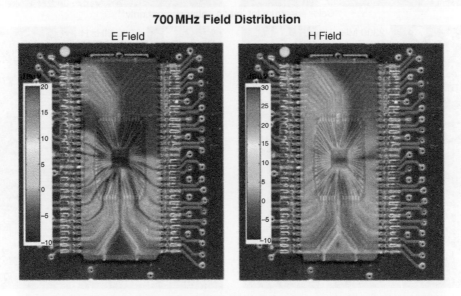

Figure 4.15: Electric and magnetic field distributions. Amplitude is in dBµV.

employ some special analytical processes that provide us with a further sharpening of emission source location. These are truly pretty pictures, with the added benefit of providing detailed information on field distributions within the package lead frame and within the silicon.

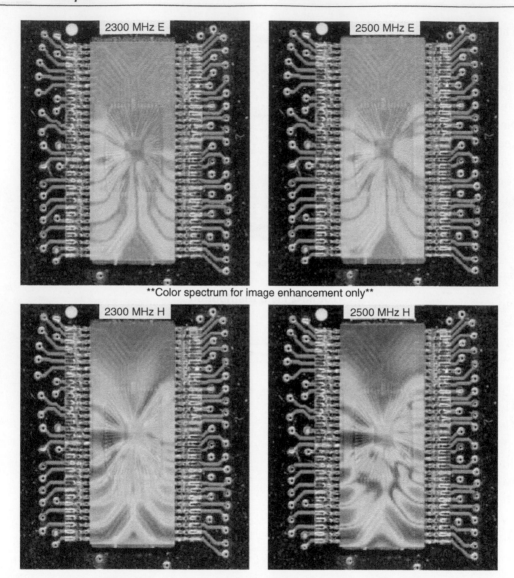

Figure 4.16: Near-field scans of the surface electric and magnetic field. These scans can show details of silicon excitation.

As the authors are employed by a semiconductor manufacturer, the ability to get devices without the encapsulation material may vary with different investigators. We do, however, encourage system designers to obtain such parts so as to better understand how these parts will impact the system platforms they are designing.

The CK505 is a device that measures 7 mm wide by 18 mm long. The die is only 1.8 mm square, which is quite small. Normally we would not expect to see much radiation from such a small device, yet there it is. We need to remember that we're not trying to communicate with satellites with the device. It is important that the device can produce emissions that have an impact in the near-field. We have to keep telling ourselves that our platforms are going to keep compressing, and this means noise sources and receptors will keep getting closer. With that in mind, we need to reject the prejudice that makes us consider only those structures that have an antenna-like appearance, or have a size commensurate with the frequencies that concern us. In close proximity, even very bad antennas will couple energy into our radios; 10 pW of power corresponds to -80 dBm, which is above the threshold for many radio receivers, as shown in Chapter 1.

4.3 Measurements Beyond the Near Field: Transition Region

Figure 4.17 is a representation of the near field, the far field, and the transition region between them. We saw earlier in the chapter that there are several methods of measuring the emissions from sources, and that these methods corresponded to the reactive near field, to the radiative far field, and to the transition region that points where energy is transferred from the near field to the far field.

The near field is a region close to the radiating source where energy is stored reactively. The energy is circulating in this region. Some energy is returning to the source, and some is continuing outward. The transition region is where the radiated energy from the source is directed more outward than inward. In the far field, the energy is almost entirely directed away from the source.

4.4 The Near-Field, Far-Field Transition

It is important to understand what we mean by the *far field* and *near field,* how we determine the transition, and what characterizes the transition. The following plots show the transition point as a function of frequency of the radiating source. Figure 4.18 shows a combined graph of the functions $1/r$, $1/r^2$, and $1/r^3$ for r between 0.1 and 2.0.

The $1/r$ term is the far-field component; $1/r^2$ and $1/r^3$ are the near-field components. The point on the plot where the three functions intersect is usually considered the crossover from near-field to far-field behavior. We will soon derive the near-field and far-field equations for a radiating dipole.

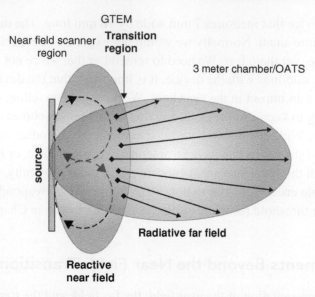

Figure 4.17: Energy flow in the near-field and far-field of a dipole radiator.

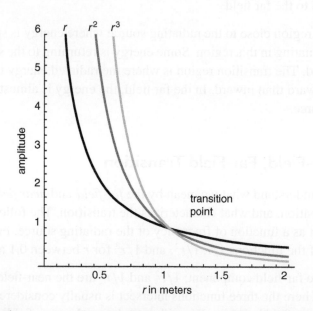

Figure 4.18: The transition from near field to far field.

Figure 4.19 shows a plot of the three functions again, but the functions have been modified with a frequency component.

$$\frac{1}{\beta_0 r}, \frac{1}{\beta_0^2 r^2}, \frac{1}{\beta_0^3 r^3} \quad \text{where} \quad \beta_0 = 2\pi/\lambda \quad \text{and} \quad \lambda = c/\text{frequency} \quad \text{and} \quad c = \text{speed of light}$$

$$c = \frac{1}{\sqrt{\mu_0 \varepsilon_0}} \quad \sqrt{\frac{\mu_0}{\varepsilon_0}} = \eta_0 \cong 377\,\Omega$$

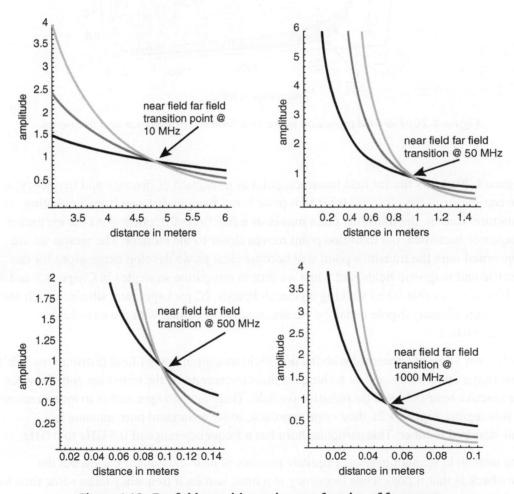

Figure 4.19: Far-field transition point as a function of frequency.

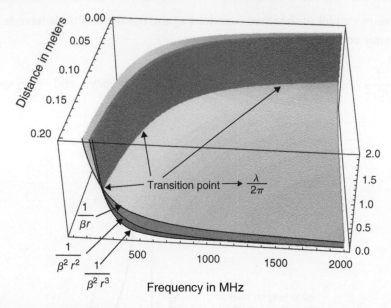

Figure 4.20: Far-field transition point as a function of distance and frequency.

Figure 4.20 shows the far-field transition point as a function of distance and frequency. As we can see from the plots, the transition point is not fixed in distance from a radiating structure. Rather, the transition point moves as a function of frequency. As the excitation frequency increases, the transition point moves closer to the radiator. The reason we are concerned with the transition point will become clear as we develop expressions for the electric and magnetic fields, and when we turn to mitigation strategies in Chapters 7 and 8. Suffice it to say that when working on circuit boards, IC packages, and silicon, which are composed of many dipole radiators, we are usually concerned with the near-field characteristics.

So far, we've been concerned with the near-field measurements of field distributions. We'll now move on to measurement techniques that concentrate on the transition zone between the reactive near field and the radiative far field. There are two approaches to measurements in this region. Figure 4.21 shows one approach, using a standard horn antenna in a non-standard manner. This particular horn has a frequency range of 0.8 GHz to 6 GHz.

The near-field scanner gives us exquisite pictures of near-field distributions, but the drawback is that it does it one frequency at a time, and each frequency takes some time to gather (depending on the spatial resolution required). The broadband horn antenna allows

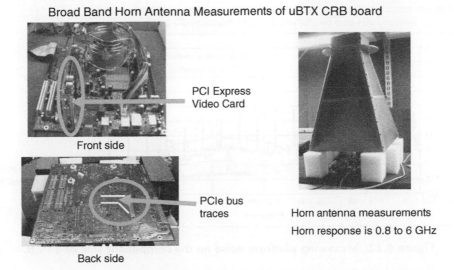

Broad Band Horn Antenna Measurements of uBTX CRB board

PCI Express
Video Card

Front side

PCIe bus
traces

Horn antenna measurements

Horn response is 0.8 to 6 GHz

Back side

Figure 4.21: Measuring platform broadband emissions from PCBs.

us to take a wide frequency snapshot of the device under test (DUT). Figure 4.22 shows a notebook motherboard that is approximately equal to the aperture of the horn, thereby giving us a nice tight coupling mechanism. This measurement method can be performed outside of a shielded or anechoic room and can be performed on the lab bench top. This method allows us to measure one side of a motherboard and isolate the other at the same time. The method also allows us to measure a wide frequency range in a short time. If we couple this method with near-field scanning, we can gather quite a bit of information at various levels of detail. The example shown in Figure 4.22 compares the emissions between two different versions of graphics programs.

In Figure 4.23, we show the GTEM described in Chapter 2. The GTEM is a quite useful device that can fit on a lab bench top and is applicable to frequencies up to 7 GHz. It can be used beyond 7 GHz, but only in a comparative manner, and it cannot be correlated to far-field measurement when used above 7 GHz. The VLSI GTEM has been used for a number of studies aimed at developing an understanding of emission processes in silicon and in their associated packages. In addition, the GTEM has been used to study high-speed differential signaling, to understand the impact of intra-pair skew in differential signaling, and to thereby develop design guidelines for high-speed system interconnects, such as the PCI Express Gen 2.

Maximum emissions from the **front side** of the board when the board is running
3dmark2001 see Benchmark (medium intensity) 3D graphics

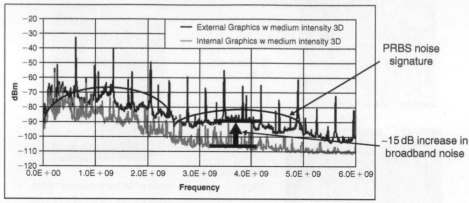

Figure 4.22: Measuring platform noise on the component side of a PCB.

VLSI GTEM: 7 GHz
BW

Test board

Device under test
is placed in the
wall of the cell

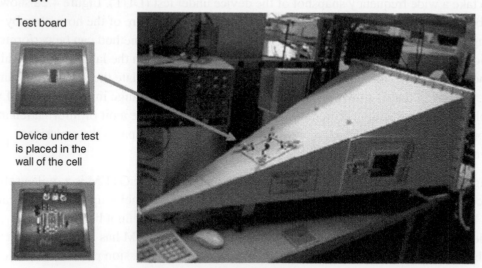

Specifically designed to measure emissions of VLSI devices to
7 GHz

Figure 4.23: VLSI GTEM: broadband measurements of system components.

We discussed some of this in Chapter 3 when we showed the impact of skewed differential signals on wireless communication channels. The GTEM approach to design development gives the designer the flexibility to investigate these potential problems quite early in the design cycle. That's the power of the method: Providing design insight early in the cycle and thereby preventing multiple design re-dos and reducing time to market considerations.

What we can see with the GTEM comparison of the radiated emissions is the significant variation across the nine devices measured. The GTEM measurements are stable and extremely repeatable. Because of the nature of the measurement, we can expect repeatability to within 1 dB. In Figure 4.24, we see variations in the emissions, even though each device passes the required functional tests. As noted, the variations are entirely due to differences in the silicon of each device.

Recall that for the most part EMI/EMC engineers generally operate from the idea of the symmetric trapezoidal signal spectrum, as shown in Figure 4.25. This signal, considered by itself, has two spectral breakpoints that are functions of pulse width and edge rate.

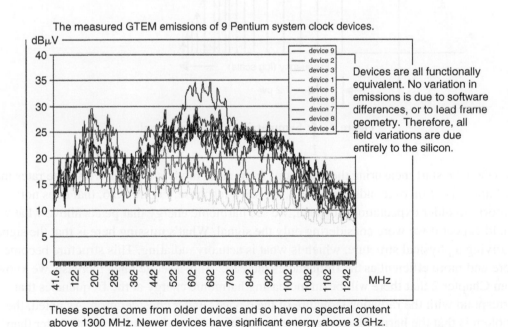

Figure 4.24: Comparison of radiated emissions from system clock devices.

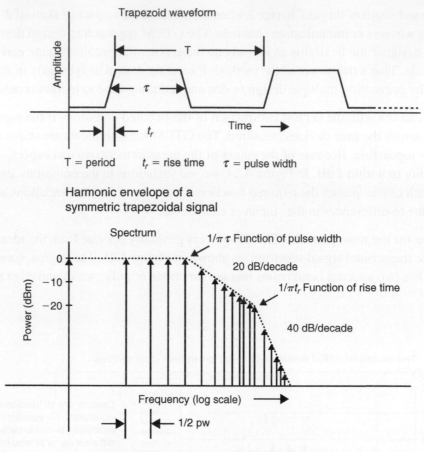

Harmonic envelope of a
symmetric trapezoidal signal

Figure 4.25: Symmetric trapezoidal signal.

So, when we start measuring the emissions from more modern devices with edge rates in the hundreds of picoseconds (Fig. 4.26), we see a response, Figure 4.26, that does not conform to older expectations. Instead, we see harmonic energy that peaks above what we would expect if we were considering only the signal. What's missing here is that the signal is driving a physical structure, which is what is actually radiating. This structure becomes more and more efficient as the frequencies increase. And, as edge rates increase, we know from Chapter 2 that there will be an increasing harmonic energy at the frequencies that correspond with the radiation structure of the devices under consideration. As noted, the problem is that the harmonic energy of the excitation source is not falling off faster than the radiation efficiency of the structure is increasing.

CK410: a more recent device
Electric field measurement ~ 3 cm off the surface of the device

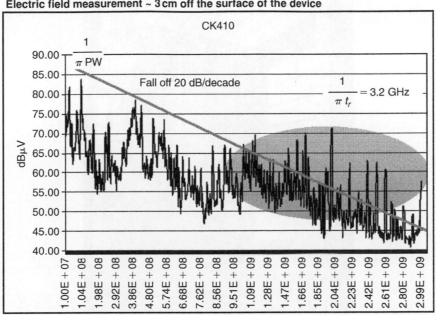

Figure 4.26: Measured emissions from a PC system clock device.

4.5 Far-Field Measurements

We will now discuss the usual EMI/EMC measurement method (the far field) performed in a semi-anechoic chamber (SAC) or at an open area test site (OATS). This is shown in Figure 4.27. Most of the authors' work has been performed in semi-anechoic chambers. Depending on the size of the chamber, one can show good correlation between an OATS and a SAC. We won't spend too much time with this type of measurement. The infrastructure is expensive and really not necessary for our purpose if we can show ultimate correlation between our bench level measurements and the OATS and SAC.

Since our prime concern has been the effect of platform sources on radio receivers, we usually are not concerned with what's going on at 10 meters, 3 meters, or even 1 meter because our platforms will almost all certainly fit within a volume much smaller than that. So why worry about far-field emissions? Well, mostly to give ourselves a certain amount of technical comfort that we know what we're doing and where we're going. A great deal

Figure 4.27: Semi-anechoic chamber. Photo courtesy of ETS-Lindgren.

of work has been done at 1 to 10 meter distances. We can use this work as calibration markers as we move in closer to our sources and receivers.

We will explore this later in a later chapter as we develop models of the near field and then use them to build models of the far field and thus correlate our work between the near field and far field.

Traditionally, there has not been much overlap between these two radiation regions. This work seeks to fill in that missing link.

4.6 Other Measurement Methods

Figure 4.29 is an example of a desktop reverberation chamber. Reverb chambers have been seeing wider use in recent years in both susceptibility and emissions testing. In susceptibility testing, they provide designers with a practical and economic means to generate high field levels without necessitating the purchase of extremely expensive power amplifiers. In addition, the mechanism of the reverb chamber creates a uniform field within

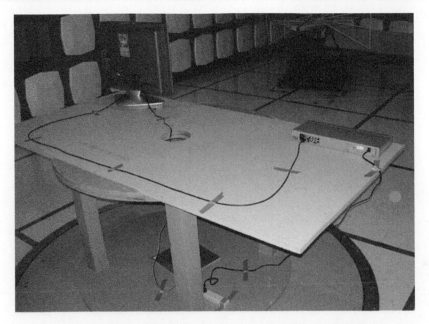

Figure 4.28: Device under test on a turn-table in the SAC.

the test space. This guarantees the user that the DUT will see a field of a given intensity at every point of the device, and that the DUT will be subjected to that interference field from every direction. The advantage to emissions testing derives from the same mechanism. The DUT is considered to be an isotropic source. Therefore, the reverb chamber can see and measure any emissions, from whatever direction they emanate. In most test situations when the user measures emissions from a DUT in a SAC or at an OATS, the device must be rotated through several angles to ensure that all emission angles have been covered.

Figure 4.30 shows the inside of the chamber. A set of rotating paddles stir the standing wave modes of the chamber, effectively spreading narrow-band energy into wider bands. As the paddles rotate, the standing wave patterns change in response, moving spatially inside the chamber region. This spatial movement of the resonances ensures that every spatial point inside the chamber will see a corresponding high field level.

One factor that all of these measurement techniques have in common is that we can build rather simple analytical models that can mimic them. Each method (except for the anechoic chamber) has the advantage of being relatively inexpensive in both money and space requirements, and each method can be shown to correlate with the others. In other words, we have a set of correlative measurements that allow us to measure and

Figure 4.29: Desk-top reverberation chamber.

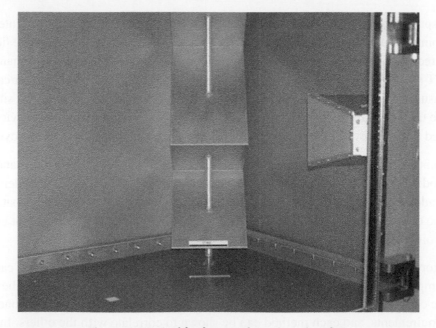

Figure 4.30: Inside the reverberation chamber.

characterize our devices from a number of different vantage points and to analyze our devices analytically with the knowledge that we can match the measurement model. We will now move on to these analytical methods.

4.7 Summary

In this chapter, we have described some of the methods that can be employed in developing mitigation methods. The given methods are by no means all of the methods that exist. The authors have found through their experience that the given methods work well together and provide a good methodology toward RFI and EMI mitigation.

The use of broadband measurement, such as the GTEM, in conjunction with a method to localize and identify emissions sources, such as near-field scanning, give the practicing engineer a nearly complete means for identifying and characterizing the interference potentials of a given platform. Neither of these methods requires the existence of expensive dedicated chambers. The GTEM can be used in an open lab environment, as can the near-field scanner. The use of a shielded room always helps, especially when there are wireless networks present, as is usually the case in large corporate environments.

When considering costs, both systems together would cost in the range of $110 K.

GTEM:	$15 K
Near-field scanner:	$30 K
Spectrum analyzer:	$45 K
Wideband preamp:	$20 K

References

[1] F. Melia, *Electrodynamics*, University of Chicago Press, 2001. (Nicely done presentation; the physics is developed along with the mathematics.)

[2] R.P. Feynman, R.B. Leighton, and M. Sands, *Feynman Lectures on Physics*, vol. 2, Addison-Wesley, 2006. (A must-have on every engineer's bookshelf.)

[3] W.A. Blanpied, *Modern Physics: An Introduction to its Mathematical Language*, Holt, Rinehart and Winston, 1971.

[4] J. Vanderlinde, *Classical Electromagnetic Theory*, 2nd ed., Springer, 2004.

[5] R.E. Hummel, *Electronic Properties of Materials*, 2nd ed., Springer-Verlag, 1993.

[6] C. Paul, *Introduction to Electromagnetic Compatibility*, Wiley, 1992.

[7] C. Paul and S. Nasar, *Introduction to Electromagnetic Fields*, McGraw-Hill, 1987. (Both Clayton Paul books are essential references that should be close at hand.)

[8] S. Ben Dhia, M. Ramdani, and E. Sicard, *Electromagnetic Compatibility of Integrated Circuits*, Springer, 2006.

[9] A. Taflove and S.C. Hagtness, *Computational Electrodynamics*, Artech House, 2005.

[10] B. Sklar, *Digital Communications*, Prentice Hall PTR, 2001.

[11] J.D. Jackson, *Classical Electrodynamics*, Wiley, 1999.

[12] K.A. Milton and J. Schwinger, *Electromagnetic Radiation: Variational Methods, Waveguides and Accelerators*, Springer, 2006.

[13] Jerry P. Marion, *Classical Dynamics of Particles and Systems*, Academic Press, 1965.

[14] K. Slattery, J. Muccioli, and T. North, *Constructing the Lagrangian of VLSI Devices from Near Field Measurements of the Electric and Magnetic Fields*, IEEE 2000 EMC Symposium, Washington, DC.

[15] K. Slattery and W. Cui, Measuring *the Electric and Magnetic Near Fields in VLSI Devices*, IEEE 1999 EMC Symposium, Seattle.

[16] K. Slattery, X. Dong, K. Daniel, *Measurement of a Point Source Radiator Using Magnetic and Electric Probes and Application to Silicon Design of Clock Devices*, IEEE 2007 EMC Symposium, Honolulu.

[17] K. Slattery, J. Muccioli, T. North, *Modeling the Radiated Emissions from Micro-processors and Other VLSI Devices*, IEEE 2000 EMC Symposium, Washington, DC.

CHAPTER 5

Electromagnetics

5.1 He Must Go by Another Way Who Would Escape This Wilderness

An antenna consists of an array of conductors that radiate energy when connected to a signal source operating at an appropriate frequency. When the antenna is excited, a current is distributed over all of the conductive surfaces. This current is determined solely by the conductor geometry and the frequency of excitation. The problem of determining the field radiated by a given current distribution is, essentially, a purely mathematical one. Unfortunately, the matter is seldom simple; it is first necessary to find the current distribution function, which is usually far more difficult mathematically. However, it is frequently possible to make a good estimate of the current distribution function, and further, to show that the finer structure of the distribution is not of prime importance. In some cases, it may be easier to estimate the local field produced by an antenna and evaluate the distant field from this. We will be concerned almost entirely with local fields. As an aside, it should be noted that the metallic materials that usually are associated with antennas are not necessary for radiation to occur. The conductors act as guides for the radiating currents, constraining the charges to flow in certain pre-determined patterns. These charge flow patterns are directly related to the radiated field patterns. One could have a beam of charges flowing in a vacuum and still have electromagnetic radiation. An example of this form of radiation is seen in storage rings used in high-energy experiments. We will not discuss this type of radiation any further.

The use of electromagnetic units can become quite confusing; at least it has been so for us. The difficulty usually is associated with factors of 4π and the speed of light, c.

Maxwell's Equations in differential form, SI units:

$$\hat{\nabla}\cdot\hat{\mathbf{E}} = \frac{\rho}{\varepsilon_0} \qquad \hat{\nabla}\cdot\hat{\mathbf{B}} = 0$$

$$\hat{\nabla}\times\hat{\mathbf{E}} = -\frac{\partial\hat{\mathbf{B}}}{\partial t} \qquad \hat{\nabla}\times\hat{\mathbf{B}} = \mu_0\hat{\mathbf{J}} + \mu_0\varepsilon_0\frac{\partial\hat{\mathbf{E}}}{\partial t} \qquad \Big\} \text{ Maxwell}$$

$$\hat{\mathbf{F}} = q(\hat{\mathbf{E}} + \hat{\mathbf{J}}\times\hat{\mathbf{B}}) \qquad \Big\} \text{ Lorentz}$$

"It is often said that the theory of electrodynamics is beautiful because of its completeness and precision. It should be noted more frequently that its appeal is indeed a measure of the elegance with which such an elaborate theoretical superstructure rests firmly on a selection of physical laws that are directly derivable from a few simple observational facts."

Figure 5.1: Maxwell's Equations.

Figure 5.1 and the equations that follow show the units between three different systems: SI, Heaviside–Lorentz, and Gaussian. (For more on this, see Milton and Schwinger.)

$$\nabla \cdot D = k_1\rho$$

$$\nabla \cdot B = 0$$

$$\nabla \times H = k_2\frac{\partial D}{\partial t} + k_1 k_2 J$$

$$\nabla \times E = -k_2\frac{\partial B}{\partial t}$$

$$D = k_3 E + k_1 P \quad \text{and} \quad H = k_4 B + k_1 M$$

$$F = q(E + k_2 v \times B) \qquad \text{Lorentz force law}$$

constant	SI	Heaviside–Lorentz	Gaussian
k_1	1	1	4π
k_2	1	$\frac{1}{c}$	$\frac{1}{c}$
k_3	ε_0	1	1
k_4	$\frac{1}{\mu_0}$	1	1
$J = \sigma E$			

where M is the magnetic material polarization, P is the electric material polarization, σ is the conductivity, and

$$c = \frac{1}{\sqrt{\mu_0 \varepsilon_0}} \qquad \sqrt{\frac{\mu_0}{\varepsilon_0}} = \eta_0 \cong 377\Omega$$

$$\mu_0 = 4\pi \times 10^{-7} \frac{\text{kilogram}^* \text{meters}}{\text{coulombs}^2} = 1.257 \times 10^{-6} \frac{\text{henry}}{\text{meter}}$$

$$\varepsilon_0 = 8.854 \times 10^{-12} \frac{\text{coulomb}^{2*} \text{seconds}^2}{\text{kilogram}^* \text{meter}^3} = 8.854 \times 10^{-12} \frac{\text{farad}}{\text{meter}}$$

$$\sigma_{DC} = 5.9 \times 10^{-7} \frac{\text{siemens}}{\text{meter}} \text{(copper)}$$

Electromagnetic equations involve the repeated use of the following mathematical symbols, as shown in Figure 5.2:

$$\nabla \qquad \text{gradient}$$

$$\nabla \cdot \qquad \text{divergence}$$

$$\nabla \times \qquad \text{curl}$$

These are differential operators. They operate on quantities to the right of them in the following ways.

∇ transforms a scalar field (voltage, temperature, and so on) into a vector field.

$\nabla \cdot$ transforms a vector field into a scalar field.

The basic operators:

$$\hat{\nabla} \equiv \frac{\partial}{\partial x}\hat{\mathbf{i}} + \frac{\partial}{\partial y}\hat{\mathbf{j}} + \frac{\partial}{\partial z}\hat{\mathbf{k}} \implies$$

An operator that acts on a scalar and produces a **vector**: **Gradient**

$$\hat{\nabla} \cdot \equiv \frac{\partial}{\partial x} + \frac{\partial}{\partial y} + \frac{\partial}{\partial z} \implies$$

An operator that acts on a vector and produces a **scalar**: **Divergence**

$$\hat{\nabla} \times \equiv \begin{vmatrix} \hat{\mathbf{i}} & \hat{\mathbf{j}} & \hat{\mathbf{k}} \\ \frac{\partial}{\partial x} & \frac{\partial}{\partial y} & \frac{\partial}{\partial z} \\ A_x & A_y & A_z \end{vmatrix} \implies$$

An operator that acts on a vector field and produces a vector field normal to the surface bounded by **C** containing vector field **A**: **Curl**

$$\nabla^2 \equiv \nabla \cdot \nabla \equiv \frac{\partial^2}{\partial^2 x} + \frac{\partial^2}{\partial^2 y} + \frac{\partial^2}{\partial^2 z} \implies$$

Divergence of the gradient which is a scalar: **Laplacian**

Figure 5.2: Analytical surface operators used with the near-field data.

$\nabla \times$ transforms one vector field into another vector field.

$$\text{Note: } \nabla \equiv \frac{\partial}{\partial x}\,\vec{i} + \frac{\partial}{\partial y}\,\vec{j} + \frac{\partial}{\partial z}\,\vec{k}.$$

The curl, $\nabla \times$, characterizes how much the field it is operating upon departs from being a conservative field. Recall that any vector field whose line integral vanishes for all closed paths is a conservative field. As an example, consider the DC voltage distributed across a resistive plane. A measurement made between any two points in the plane is dependent only upon the two measurement points and not on the particular path connecting them. This would not in general be true for an AC voltage distributed across the same resistive plane.

When the curl of a quantity is not equal to zero, this is saying that it is path dependent.

Let's take the equations in sets and arrange them in logical order. In Figure 5.3, we group the electrostatic equations together and the magnetostatic equations together. We also point out the symmetry of this ordering by noting the complementarity of the zero elements and the charge and charge density elements. As these are the static equations, it should not be surprising that there aren't any time-dependent terms.

This might be the moment to make an observation: How do the charges know that it is a time-independent situation? How is it known that nothing is changing? We need to imagine the structure that supports our configuration of charges. In Figure 5.3, we indicate only two distinct charges, but it can be generalized to any number of charges. A force is exerted between q_1 and q_2 directed along the line between the two charges and is directly

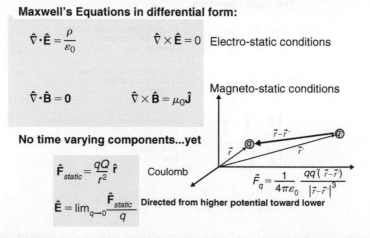

Maxwell's Equations in differential form:

$$\hat{\nabla} \cdot \hat{\mathbf{E}} = \frac{\rho}{\varepsilon_0} \qquad \hat{\nabla} \times \hat{\mathbf{E}} = 0 \quad \text{Electro-static conditions}$$

$$\hat{\nabla} \cdot \hat{\mathbf{B}} = 0 \qquad \hat{\nabla} \times \hat{\mathbf{B}} = \mu_0 \hat{\mathbf{J}}$$

Magneto-static conditions

No time varying components...yet

$$\hat{\mathbf{F}}_{static} = \frac{qQ}{r^2}\,\hat{\mathbf{r}} \quad \text{Coulomb}$$

$$\hat{\mathbf{E}} = \lim_{q \to 0} \frac{\hat{\mathbf{F}}_{static}}{q} \quad \text{Directed from higher potential toward lower}$$

$$\vec{F}_q = \frac{1}{4\pi\varepsilon_0} \frac{qq'\,(\vec{r}-\vec{r}')}{|\vec{r}-\vec{r}'|^3}$$

Figure 5.3: Maxwell's Equations: the static conditions.

proportional to the product of the charges involved and inversely proportional to the square of the distance between them. The resulting electric field is related to the force between the charges. But, how does charge 1 know that charge 2 is anywhere nearby? How does this information get from one charge to the other? For the electromagnetic field, there is a carrier of field information called the *photon*. Electric charge will absorb and emit these photons. In Figures 5.4 to 5.7, we will look more closely at the electrostatic conditions.

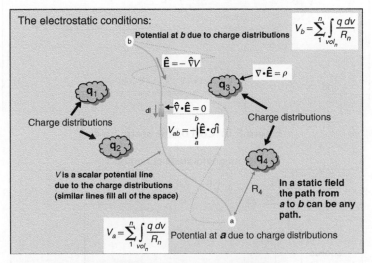

Figure 5.4: A scalar potential line.

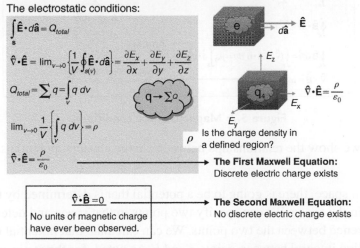

Figure 5.5: Maxwell's First and Second Equations.

Curl of the electro-static field

$V_{ab} = \int_a^{b \to a} \hat{\mathbf{E}} \cdot d\hat{\mathbf{l}} = 0$ If C is a closed curve, and the electric field is static (called "conservative") then = 0

$\oint_c \hat{\mathbf{E}} \cdot d\hat{\mathbf{l}} = \oint_s (\hat{\nabla} \times \hat{\mathbf{E}}) \cdot \hat{n} \, da$ Stokes

Where n is the unit vector normal to the surface S

C is the closed loop
S is the surface bounded by *C*

and $(\hat{\nabla} \times \hat{\mathbf{E}}) = 0$ ———→ **Maxwell's Third Equation for the electro-static condition**

Curl of E

Figure 5.6: Curl of the static field.

The magneto-static condition

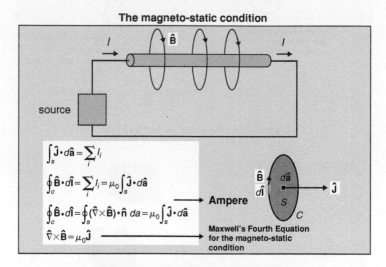

$\int_s \hat{\mathbf{J}} \cdot d\hat{\mathbf{a}} = \sum_i I_i$

$\oint_c \hat{\mathbf{B}} \cdot d\hat{\mathbf{l}} = \sum_i I_i = \mu_0 \int_s \hat{\mathbf{J}} \cdot d\hat{\mathbf{a}}$

$\oint_c \hat{\mathbf{B}} \cdot d\hat{\mathbf{l}} = \oint_s (\hat{\nabla} \times \hat{\mathbf{B}}) \cdot \hat{n} \, da = \mu_0 \int_s \hat{\mathbf{J}} \cdot d\hat{\mathbf{a}}$ ——→ **Ampere**

$\hat{\nabla} \times \hat{\mathbf{B}} = \mu_0 \hat{\mathbf{J}}$ ——————→ **Maxwell's Fourth Equation for the magneto-static condition**

Figure 5.7: Magneto-static conditions.

In Figure 5.4, we show the relationships between charge clusters, potentials, voltage, and electric field.

At each point in space, there is going to be a potential that is determined by the distribution of all charges in the space. We can specify two points in that space and determine the potential difference between the two points. We can calculate that potential difference with the use of the line integral between points *a* and *b*. As noted, for the static condition, it doesn't matter where we place the line connecting the two points; we get the same answer no matter how convoluted the line we draw.

We will now look at the equations a little more closely. In Figure 5.5, we show that the dot product of the electric field over a general closed surface is equal to the total amount of charge contained within the surface. The charge density in a volume is equal to the total charge within the volume and is also equal to the dot product of the differential operator and the electric field. This is Maxwell's First Equation, which states that a discrete electric charge exists because we can keep shrinking the volume-enclosing charges until the volume encloses a single piece of charge. Note also that we can take the dot product of the differential operator and the magnetic field and shrink the volume down to nothing and never enclose a single piece of magnetic charge. No equivalent discrete magnetic charge exists similar to electric charge. This is Maxwell's Second Equation. However, there are occasions when we can assume that magnetic charge exits in order to simplify the mathematics on our way to a particular solution.

We've looked at the *divergence* of the electric and magnetic fields. We'll now look at the *curl* of the fields. We consider the case where our line of field integration closes on itself and forms a closed curve. In this case, the resulting potential is equal to zero. This is the case for a static electric field. A field that doesn't change with time is called *conservative*. Using Stokes Theorem, we can then equate the curl of the electric field with the dot product of the electric field and the line element of the closed curve and see that the curl is also equal to zero. This is shown in Figure 5.6 and illustrates Maxwell's Third Equation: that the curl of the electric field is equal to zero.

When we get to the time-varying case, we'll see that it is no longer the case that the path of integration is unimportant. The integral will vary as a function of the particular path that we take. We will also see that the integral is not zero when we curve it back on itself.

For the electrostatic condition, we need a static assembly of charges. For the magnetostatic case, we need a moving set of charges. Figure 5.7 shows a constant current, which is a set of charges in non-accelerated motion. If we were to draw a surface that bisected the conductor where the charges were located, we could take the integral of the charge density dotted with the elemental unit of surface area and get the total charge crossing that surface in a given amount of time. We would also equate the line integral of the magnetic field around the conductor with the integral of the charge density. We can then use Stokes' Theorem to equate the line integral of the magnetic field with the curl of the magnetic field through the surface defined by the curve. This is Ampere's Law and is Maxwell's Fourth Equation for electrostatic conditions.

5.2 The Time-Varying Maxwell's Equations

We've been looking at the static conditions for Maxwell's Equations, so at this point we can ask a question: How does a capacitor work? We asked this question a little differently earlier. Now we're asking again: How does one plate of a capacitor know the state of the charges on the other plate of the capacitor? When we finish with the time-varying conditions, we should be in a position to answer this question.

In Figure 5.8, we show the case of a magnetic field changing with time and intercepting a conductor loop. We can see that the line integral around the loop is not equal to zero but rather is a voltage that is varying with time similar to the magnetic field crossing the surface defined by the wire loop. Instead of the curl of the electric field being equal to zero, it is now proportional to the partial derivative with respect to time of the magnetic field (which is also proportional to the line integral of the electric field along the curve, C).

As noted in Figure 5.8, if you measure the voltage in the electrostatic condition, the arrangement of the probe leads does not factor in or have an effect on the measurement. However, when the measurement is of a time-varying signal, the probe lead arrangement does make a difference. In any measurement, we need to account for the impact of the measurement system itself.

In Figure 5.9, we show the conditions for a time-varying current propagating in a conductor and the space between two capacitor plates. As noted, if we have only the pieces

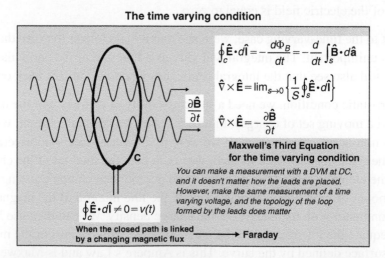

The time varying condition

$$\oint_c \hat{\mathbf{E}} \cdot d\hat{\mathbf{l}} = -\frac{d\Phi_B}{dt} = -\frac{d}{dt}\int_s \hat{\mathbf{B}} \cdot d\hat{\mathbf{a}}$$

$$\hat{\nabla} \times \hat{\mathbf{E}} = \lim_{s \to 0}\left\{\frac{1}{S}\oint_s \hat{\mathbf{E}} \cdot d\hat{\mathbf{l}}\right\}$$

$$\hat{\nabla} \times \hat{\mathbf{E}} = -\frac{\partial \hat{\mathbf{B}}}{\partial t}$$

**Maxwell's Third Equation
for the time varying condition**

You can make a measurement with a DVM at DC, and it doesn't matter how the leads are placed. However, make the same measurement of a time varying voltage, and the topology of the loop formed by the leads does matter

$\dfrac{\partial \hat{\mathbf{B}}}{\partial t}$

$$\oint_c \hat{\mathbf{E}} \cdot d\hat{\mathbf{l}} \neq 0 = v(t)$$

**When the closed path is linked
by a changing magnetic flux** ⟶ **Faraday**

Figure 5.8: The time-varying condition: Maxwell's Third Equation.

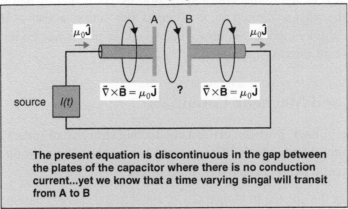

Figure 5.9: Making the equations consistent.

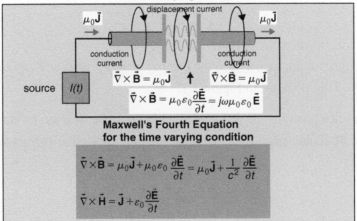

Figure 5.10: Maxwell's Fourth Equation for the time-varying condition.

of Maxwell's Equations so far mentioned, then we have a discontinuous situation across the plates. How do we enforce the continuity of charge across the gap?

In Figure 5.10, we add the piece that Maxwell added to the equations to make everything work. The added piece is the displacement current. In Maxwell's time, scientists knew that current could be composed of real electric charges flowing in metal conductors, but they had no idea of photons. The use of displacement current as a current component across the gap in a capacitor is an interesting addition for Maxwell to make. He had some mechanical models to describe what was going on, as all good Victorian scientists of the time had, but they were rather clunky models. So, now we have all the pieces that go into constructing electrodynamics and electromagnetics.

5.3 Electric and Magnetic Potentials

As mentioned previously, if we are given a conducting structure of some type, and we can describe the current distribution in that structure, we are then in a position to find the electric and magnetic fields generated by that current.

The divergence of the electric field intensity is equal to the charge contained within the region of interest $\vec{\nabla} \cdot \varepsilon_0 \vec{E} = \rho$, where we are transforming a vector, the electric field, into a scalar, the charge. Also, $\vec{E} = \vec{\nabla} V$, where $\vec{\nabla} V$ is the gradient of the electric scalar potential. Therefore,

$$\vec{\nabla} \cdot \vec{\nabla} V = \nabla^2 V = -\frac{\rho}{\varepsilon_0}.$$

Knowing the electric field, we can find the *electric scalar potential*, V, by integrating the electric field, $\vec{E}$. This is shown in Figure 5.11.

The magnetic vector potential, $\vec{A}$ is approached in the same manner. Its curl is equal to $\mu_0 \vec{H}$ (magnetic field intensity), and it satisfies an equation similar to the electric scalar potential:

$$\nabla^2 \vec{A} = \mu_0 \vec{J}.$$

Note, however, that the quantity to the right of the differential operator is a vector, and that the right side of the equation is a vector, hence, electric *scalar* potential, and magnetic *vector* potential. One can think of electric scalar potential as relating to a *stationary **point** charge* and the magnetic vector potential relating to an extended, and directed, (***line***) *current density in motion.*

The divergence of the magnetic field intensity is zero.

$$\vec{\nabla} \cdot \mu_0 \vec{H} = 0$$

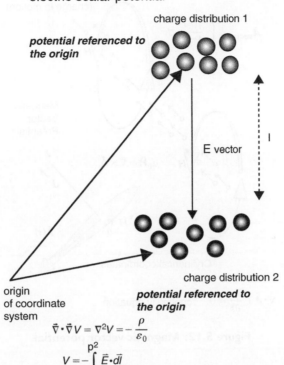

electric scalar potential

charge distribution 1

potential referenced to the origin

E vector

l

origin
of coordinate
system

charge distribution 2

potential referenced to the origin

$$\vec{\nabla} \cdot \vec{\nabla} V = \nabla^2 V = -\frac{\rho}{\varepsilon_0}$$

$$V = -\int_{p1}^{p2} \vec{E} \cdot \vec{dl}$$

Figure 5.11: The electric scalar potential.

This is valid everywhere in a region because there are no known discrete magnetic charges as there are electric charges. (The magnetic field lines close upon themselves, not on other charges as in the electric field.)

Using the vector identity relation

$$\vec{\nabla} \cdot \vec{\nabla} \times \vec{A} \equiv 0$$

we get

$$\mu_0 \vec{H} = \vec{\nabla} \times \vec{A}$$

where $\vec{A}$ is the magnetic vector potential. This is shown in Figure 5.12.

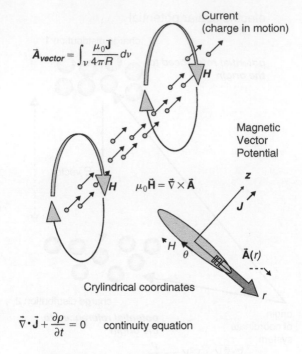

$$\vec{A}_{vector} = \int_v \frac{\mu_0 \mathbf{J}}{4\pi R} dv$$

Current
(charge in motion)

H

Magnetic
Vector
Potential

$$\mu_0 \vec{H} = \vec{\nabla} \times \vec{A}$$

z

H

J

H
θ

$\vec{A}(r)$

r

Crylindrical coordinates

$$\vec{\nabla} \cdot \vec{J} + \frac{\partial \rho}{\partial t} = 0 \quad \text{continuity equation}$$

Figure 5.12: Magnetic vector potential.

5.4 Radiation Mechanisms

Given the two charge distributions, how is the static electric field maintained between them? That is, what is the form of the information that passes between the distributions so that each "knows" about the other? Is the electric field a distortion in space that guides a charge along the path that requires the least energy? If a distortion, what is it that is being distorted? Or, perhaps, the electric field information is transmitted between the distributions by photons. Photons are emitted by accelerated charge, but in this case the charges are considered to be at rest. In order to posit the photon as transmitting the information, we must surmise that the charges are not truly at rest. This can be the case if we consider them as being in the real world, and therefore having some actual kinetic energy. Even though the distributions are constrained to a specific region on a macroscopic scale, the individual charges must be jostling within that circumscribed region. Because the charges have accelerated movement, they emit photons. These photons form the electric field between the distributions. The electric field now being seen to be a statistical effect from measuring many photons.

Figures 5.13 to 5.16 show various ideas about radiation mechanisms in metals.

A charged particle, when accelerated, will emit a photon. When charge absorbs a photon, it will in turn respond by accelerating. In this way, the signal propagates in a metal. In addition, the photon is emitted at right angles to the direction of charge acceleration.

Two mechanisms must be in action in order to have an antenna radiate. There must be charge movement in two orthogonal motions—along the axis of the antenna and orthogonal to the long axis of the antenna. The transverse motion maintains the signal propagation within the metal; the longitudinal motion contributes to the radiated energy that leaves the antenna. Figures 5.15 and 5.16 indicate these mechanisms. Figure 5.15 shows a piece of the metal conductor and illustrates the two motions and the resulting energy propagation within the metal and outward from the metal.

Figure 5.16 attempts to bring all of these elements together in a single visual representation. As noted, an individual charge (an electron) will not travel very far in the metal lattice before it collides with a metal atom and is either absorbed or scattered. In addition, the electrons do not travel anywhere near the speed of light inside the metal. Indeed, it takes huge particle accelerators using many megawatts of power in order to get something as light as an electron up near the speed of light. So, this again tells us that

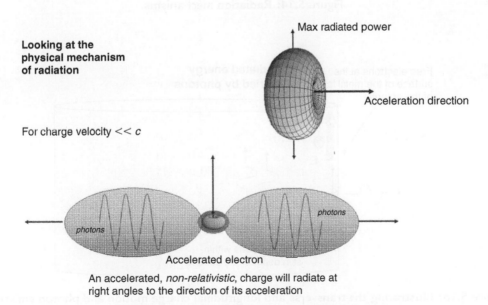

Looking at the physical mechanism of radiation

For charge velocity $\ll c$

Max radiated power

Acceleration direction

photons

photons

Accelerated electron

An accelerated, *non-relativistic*, charge will radiate at right angles to the direction of its acceleration

Figure 5.13: Radiation mechanisms.

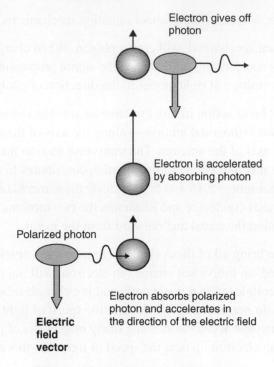

Electron gives off photon

Electron is accelerated by absorbing photon

Polarized photon

Electron absorbs polarized photon and accelerates in the direction of the electric field

Electric field vector

Figure 5.14: Radiation mechanisms.

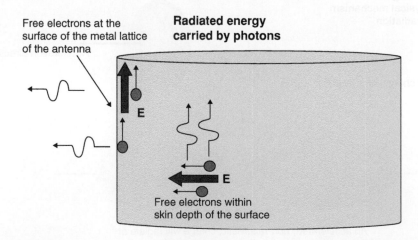

Free electrons at the surface of the metal lattice of the antenna

Radiated energy carried by photons

E

E

Free electrons within skin depth of the surface

Figure 5.15: Illustrating the transverse and longitudinal charge motion and photon emission.

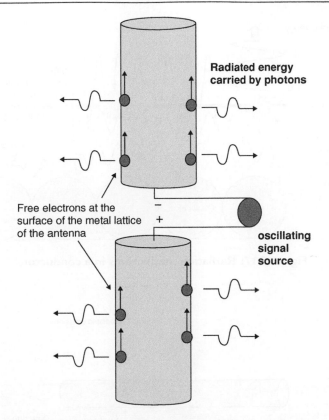

Figure 5.16: Charge motion and photon emission in a dipole antenna.

charges do not carry the information about an electromagnetic event from one place to another within a system. It must be something else that can propagate at the speed of light within the medium under consideration, and this can only be the photon.

When we consider Figure 5.17, we need to include the boundary conditions: how the electric field behaves at the metal boundary and how the magnetic field behaves at the boundary and across the boundary. We must also keep in mind that the signal event must be propagating from one end of the system to the other.

Figure 5.18 shows radiated emissions leaked from a transmission line. The figure shows the radiation mechanisms that provide the transmission of information along the direction of the line and the leakage of power from the line that becomes radiated emissions. Keep in mind that 1 pW of radiated power is enough to interfere in the operation of most radios,

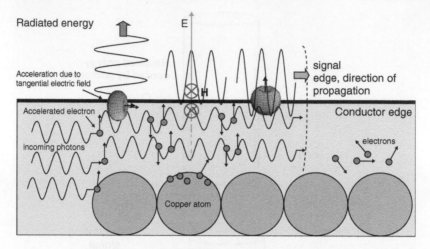

Figure 5.17: Radiation mechanisms in a conductor.

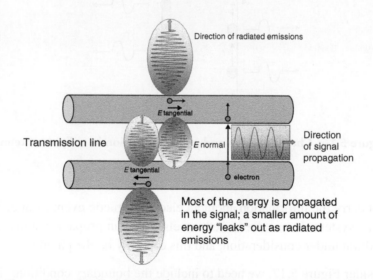

Figure 5.18: Radiated emissions leaked from a transmission line.

as mentioned in Chapter 1. So, 1 pW of leaked power would never be noticed during signal integrity considerations. In fact, good luck measuring 1 pW with the typical instruments used during signal measurement.

In the next chapter, we will start using the ideas that we have been discussing here. We will build models using Maxwell's Equations and the simple inferences from these equations.

References

[1] F. Melia, *Electrodynamics,* University of Chicago Press, 2001. (Nicely done presentation; the physics is developed along with the mathematics.)

[2] R.P. Feynman, R.B. Leighton, and M. Sands, *Feynman Lectures on Physics,* vol. 2, Addison-Wesley, 2006. (A must-have on every engineer's bookshelf.)

[3] W.A. Blanpied, *Modern Physics: An Introduction to its Mathematical Language,* Holt, Rinehart and Winston, 1971.

[4] J. Vanderlinde, *Classical Electromagnetic Theory,* 2nd ed., Springer, 2004.

[5] R.E. Hummel, *Electronic Properties of Materials,* 2nd ed., Springer-Verlag, 1993.

[6] C. Paul, *Introduction to Electromagnetic Compatibility,* Wiley, 1992.

[7] C. Paul and S. Nasar, *Introduction to Electromagnetic Fields,* McGraw-Hill, 1987. (Both Clayton Paul books are essential references that should be close at hand.)

[8] S. Ben Dhia, M. Ramdani, and E. Sicard, *Electromagnetic Compatibility of Integrated Circuits,* Springer, 2006.

[9] A. Taflove and S.C. Hagtness, *Computational Electrodynamics,* Artech House, 2005.

[10] B. Sklar, *Digital Communications,* Prentice Hall PTR, 2001.

[11] J.D. Jackson, *Classical Electrodynamics,* Wiley, 1999.

[12] K.A. Milton and J. Schwinger, *Electromagnetic Radiation: Variational Methods, Waveguides and Accelerators,* Springer, 2006.

[13] J.P. Marion, *Classical Dynamics of Particles and Systems,* Academic Press, 1965.

[14] K. Slattery, J. Muccioli, and T. North, *Constructing the Lagrangian of VLSI Devices from Near Field Measurements of the Electric and Magnetic Fields,* IEEE 2000 EMC Symposium, Washington, DC.

[15] K. Slattery and W. Cui, Measuring *the Electric and Magnetic Near Fields in VLSI Devices,* IEEE 1999 EMC Symposium, Seattle.

[16] K. Slattery, X. Dong, K. Daniel, *Measurement of a Point Source Radiator Using Magnetic and Electric Probes and Application to Silicon Design of Clock Devices,* IEEE 2007 EMC Symposium, Honolulu.

[17] K. Slattery, J. Muccioli, T. North, *Modeling the Radiated Emissions from Microprocessors and Other VLSI Devices,* IEEE 2000 EMC Symposium, Washington, DC.

CHAPTER 6

Analytical Models

6.1 Honest Analysis Gets in the Way of Results Desired Emotionally

We will begin this chapter by considering the physics of simple dipole radiators. For many applications, we can approximate the true situation quite well if we begin our analysis by reducing the problem to that of a multiple set of dipole radiators. There are occasions where more serious simulators are required, but they aren't always required, and so it pays to be able to use the simple equations of dipoles. Let's get at it.

Using models to understand the physics of interference.

6.2 The Electric Dipole

Consider an elementary Hertzian dipole, as shown in Figure 6.1. The length, dl, is assumed to be infinitesimally short. The current distribution in the dipole arm is assumed to be uniform along the length of the dipole (and this is a big simplifying assumption). From this, we can then deduce that the current vector has only a z-directed component.

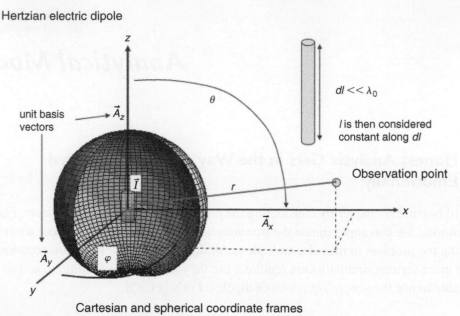

Cartesian and spherical coordinate frames

Figure 6.1: An elementary electric dipole.

Therefore, the magnetic vector potential itself also reduces to having only a z-directed component.

Note that the following explanation is quite terse and is only meant to indicate the flow of the mathematical development. The interested reader is pointed to the many excellent texts on the subject, especially Clayton Paul's *Introduction to Electromagnetic Fields* and, of course, Richard Feynman's *Lectures on Physics*, Vol. 2.

The excitation current in the dipole is assumed to be entirely in the z-direction.

$$\vec{A}_z = \frac{\mu_0}{4\pi} \vec{I} \frac{dl}{r} e^{-j\beta_0 \vec{r}} \quad \vec{A}_x = 0 \quad \vec{A}_y = 0$$

In spherical coordinates, this becomes vector $A = \{A_r, A_\theta, A_\varphi\}$. Through transformation from Cartesian to Spherical, the individual components of the vector are

$$A_r = A_z \cos\theta$$

$$A_\theta = -A_z \sin\theta$$

$$A_\varphi = 0.$$

The magnetic field can now be found by taking the curl of the magnetic vector potential,

$$\vec{H} = \frac{1}{\mu_0} \vec{\nabla} \times \vec{A}.$$

The electric field is obtained by taking the curl of the magnetic field,

$$\vec{E} = \frac{1}{j\omega\varepsilon_0} \vec{\nabla} \times \vec{H}. \tag{6.1}$$

Performing the indicated operations, and with a little algebraic work, the following equations are obtained for the $\vec{E}$ field:

$$\vec{E}_r = \frac{m_e}{2\pi} \eta_0 \beta_0^2 \cos\theta \left(\frac{1}{\beta_0^2 r^2} - j\frac{1}{\beta_0^3 r^3} \right) e^{-j\beta_0 r} \tag{6.2}$$

$$\vec{E}_\theta = \frac{m_e}{4\pi} \eta_0 \beta_0^2 \sin\theta \left(j\frac{1}{\beta_0 r^1} + \frac{1}{\beta_0^2 r^2} - j\frac{1}{\beta_0^3 r^3} \right) e^{-j\beta_0 r} \tag{6.3}$$

$$\vec{E}_\varphi = 0 \qquad \eta_0 = \sqrt{\frac{\mu_0}{\varepsilon_0}},$$

where m_e is the electric dipole moment, $\vec{I}dl$, $\beta = 2\pi/\lambda$.

To show how we obtain these equations, we'll show the algebra for the radial electric field component.

$$\frac{1}{j\omega\varepsilon_0} \vec{\nabla} \times \vec{H} \Rightarrow \vec{E}_r = \frac{m_e}{2\varepsilon_0\omega\pi r^3} \cos\theta(-j + \beta_0 r) e^{-j\beta_0 r}$$

noting that $\omega = \dfrac{\beta_0}{\sqrt{\varepsilon_0\mu_0}}$

$$\frac{m_e\sqrt{\varepsilon_0\mu_0}\cos\theta(-j + \beta_0 r)e^{-j\beta_0 r}}{2\beta_0\varepsilon_0\pi r^3} \Rightarrow \text{removing the terms we can ignore}$$

$$\frac{\sqrt{\varepsilon_0\mu_0}(-j + \beta_0 r)}{\varepsilon_0\beta_0 r^3}$$

$$\frac{\sqrt{\varepsilon_0\mu_0}}{\varepsilon_0} \left(\frac{-j}{\beta_0 r^3} + \frac{\beta_0 r}{\beta_0 r^3} \right) = \frac{\sqrt{\varepsilon_0\mu_0}}{\varepsilon_0} \left(\frac{-j}{\beta_0 r^3} + \frac{1}{r^2} \right)$$

$$\frac{\sqrt{\varepsilon_0 \mu_0}}{\varepsilon_0} = \frac{\left(\sqrt{\varepsilon_0 \mu_0}\right)^2}{\varepsilon_0^2} = \frac{\varepsilon_0 \mu_0}{\varepsilon_0^2} = \sqrt{\frac{\mu_0}{\varepsilon_0}} = \eta_0$$

$$\eta_0 \beta_0^2 \left(\frac{1}{\beta_0^2 r^2} - j \frac{1}{\beta_0^3 r^3} \right)$$

We want it in this form, and it clearly shows the near-field and far-field relations.

The associated $\vec{H}$ field is

$$\vec{H}_\varphi = \frac{m_e}{4\pi} \beta_0^2 \sin\theta \left(j \frac{1}{\beta_0 r} + \frac{1}{\beta_0^2 r^2} \right) e^{-j\beta_0 r} \tag{6.4}$$

with $\vec{H}_\theta = \vec{H}_r = 0$.

Figure 6.2 is a plot of the magnitude of the radial electric field, $\vec{E}_r$, from $-3\,\text{m}$ to $3\,\text{m}$ distance from the dipole radiator, and varying the angle theta $0 < \theta < \pi$ at a frequency of $300\,\text{MHz}$. The amplitude is in volts.

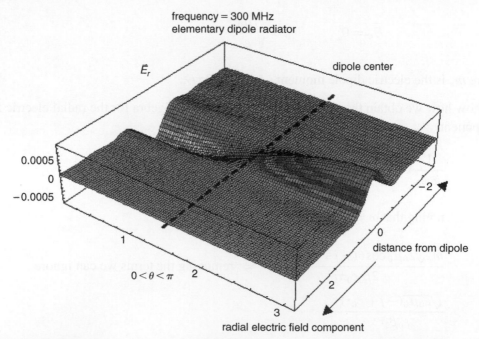

Figure 6.2: Radial component of the electric field from an elementary electric dipole: Linear plot.

We can see from this plot that the $\vec{E}$ field seen along an axis directed from the dipole center is a minimum path, and the field reaches a maximum at either end of the dipole. The radial electric field is present only in the near field and does not radiate into the far field. Because of this, it is considered to be part of the reactive near field. Energy is stored but not radiated (well, not far, at least) in the radial electric field.

Figure 6.3 is a plot of the magnitude of the $\vec{E}$ field transverse component.

As it expands outward, the field becomes concentrated toward the center of the dipole. Radiation off the ends of the dipole diminishes, and in the limit will go to zero. Consequently there isn't any energy storage at the dipole tips.

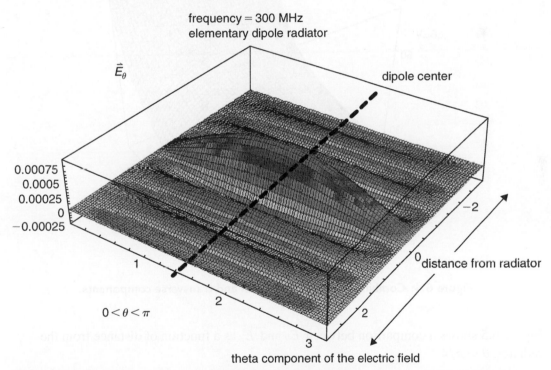

Figure 6.3: Transverse component of the electric field from an elementary electric dipole: Linear plot.

In Figure 6.4, we combine the two electric field components generated by the dipole into one response plot. As we can see, the radial component is dominant in close to the dipole and the transverse component begins to dominate at further distances from the dipole. Recall the transition region shown earlier. As mentioned earlier, for most EMC applications we are concerned only with the action of the transverse component. When we start looking into platform interference issues, we find ourselves required to deal with all the components generated by radiating sources. Therefore, we need models and analytical approaches that allow us to develop an intuitive feel for near-field behavior.

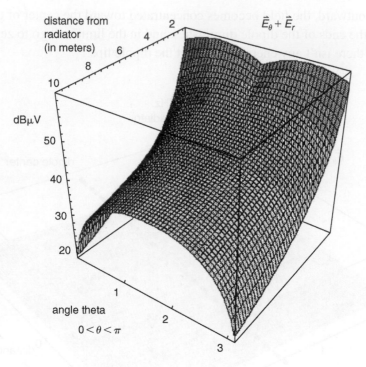

Figure 6.4: Combined effect of the radial and transverse components.

Figure 6.5 shows a comparison between $\vec{E}_\theta$ and $\vec{E}_r$ as a function of distance from the radiator. $\theta = \pi/4$

A plot of the third dipole component, the magnetic component $\vec{H}_\varphi$, would show a surface similar to that for $\vec{E}_\theta$ with maximum amplitudes differing by approximately 50 dB in the far field, and varying values in the near field. The reason for this difference in amplitude is that the electric field and the magnetic field are related by the wave impedance of free

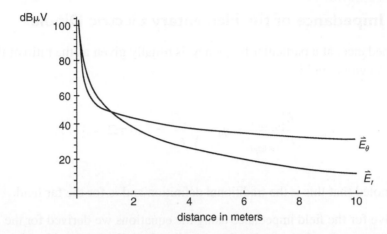

Figure 6.5: Distance comparison of radial and transverse components from the electric dipole.

space. Figure 6.6 shows a comparison between the electric and magnetic field magnitudes. Note that the field surface for the transverse electric and magnetic components is similar though differs in amplitude for the elementary dipole. We will see that this is not the case when we consider the extended dipole.

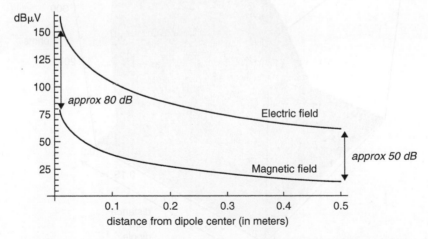

Figure 6.6: Magnitude relation between electric and magnetic fields.

6.3 Field Impedance of the Elementary Electric Dipole

The field impedance, at a particular frequency, is usually given as the ratio of the electric field and the magnetic field.

$$Z_0 = \frac{|E|_{\text{far field}}}{|H|_{\text{far field}}} = \eta_0 = \sqrt{\frac{\mu_0}{\varepsilon_0}} = 120\,\pi\Omega \approx 377\,\Omega \tag{6.5}$$

It should be noted that this is the traditional definition and is for the far field.

When we solve for the field impedance using the equations we derived for the H and E fields, we obtain a general expression that describes both the near-field and the far-field

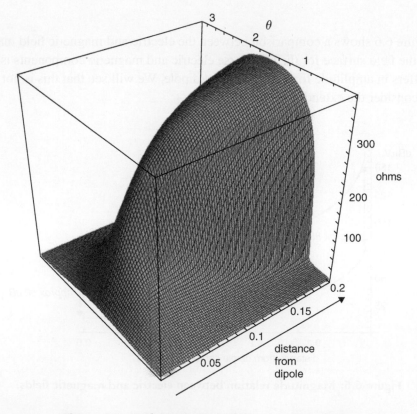

Figure 6.7: Real component of the wave impedance.

impedance. In the limit, as $r \to \infty$, we get approximately 377 ohms. The impedance of the near field for an electric dipole is

$$Z = \eta_{\text{near field}} = \sqrt{\left(\frac{E_\theta}{H_\varphi}\right)^2 + \left(\frac{E_r}{H_\varphi}\right)^2}. \tag{6.6}$$

This is the impedance of the radiated field at the *particular* chosen frequency, which, for our examples, is 300 MHz.

Figure 6.7 is a graph of the real portion of the ratio of E field and H field. The plot shows that the boundary conditions are observed: In the far field, the value is 377 ohms, and at the conductor surface it goes to zero, as we should expect. The angle theta varies as $0 < \theta < \pi$.

Figure 6.8 shows the magnitude of the wave impedance. This plot includes both real and reactive parts, $Z = |R + jX|$.

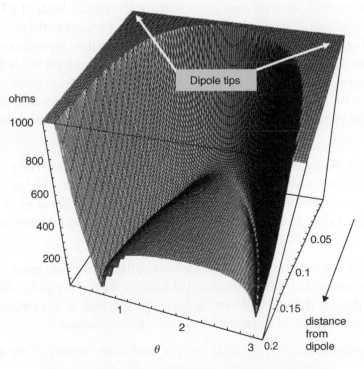

Figure 6.8: Absolute value of the field impedance.

Along an axis directed from the center and perpendicular to the dipole axis, the value of the impedance goes to 377 ohms, and at the tips the value goes to infinity. There is no power radiated axially off the tips of an electric dipole.

The point of presenting the field equations and the plots showing the field variations and the impedance variations in the near field is to indicate the complexity of measurements conducted in the near field. Also, if we use the near field scanner and observe the distributions and the relative magnitudes of the electric and magnetic field components, we may get some insight into the impedance of the radiating source and thereby be able to infer structural details that may be hidden within the silicon and the package.

Also, we should mention that the wave impedance is independent of the current intensity and is related solely to the geometry of the radiator and the frequency of excitation.

6.4 The Extended Dipole Radiator

In the previous sections, we were concerned with the theoretical fields of a Hertzian dipole, which is of infinitesimal length. This is clearly not related to the real world. However, the field equations are of rather simple form and allow for extensive treatment in an open manner. The following brief discussion will demonstrate the fields derived from a dipole of measurable length. The plots given here were derived from equations developed by R.S. Elliot in his textbook, *Antennas and Systems*. We will not develop the equations as we did for the Hertzian dipole, but we will simply present the results with some discussion.

Figure 6.9 shows the transmitted electric field of an extended dipole. The transmitted field is the portion that actually escapes the near field (and develops into the far field) and will be measured at some remote distance from the radiating device. These plots show the calculated fields in spherical coordinates—the angles of the fields in relation to the axis of the dipole.

In Figure 6.9, we can see three significant lobes of radiation: off the center of the dipole, and off the two tips of the dipole. However, the field off the tips is rapidly decreasing along the axis of the dipole. This is an interesting result. We can look at an extended radiating structure, such as a dipole, as an ensemble of three distinct radiators.

Figure 6.10 shows the reactive, or radial, electric field in the vicinity of the dipole.

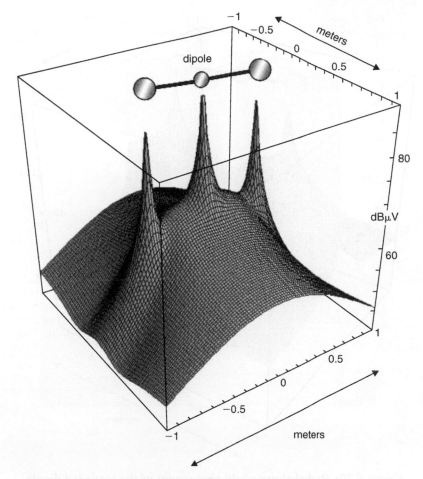

Figure 6.9: Transverse electric field component of the extended dipole.

Note in the radial plot that we have an essentially complementary situation to that of the transverse radiative. The center of the dipole is now seen to not radiate outward; the tips of the dipole are sites of maximum field strength. This is the reactive portion of the near field from the dipole radiator. No part of the reactive field makes it into the far field.

Figure 6.11 shows the radiated magnetic field of the extended dipole.

The magnetic field is seen to radiate only from the arms of the dipole and not from the tips or the center when observed close to the dipole. As we move further from the dipole, we

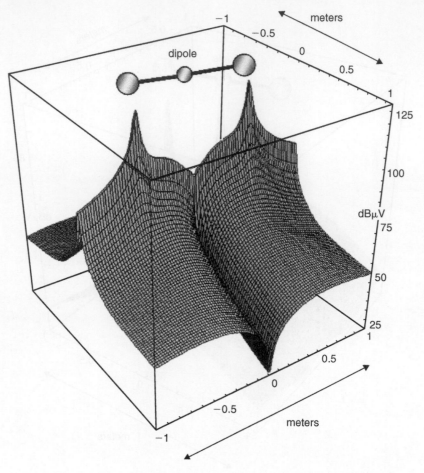

Figure 6.10: Radial electric field component of the extended dipole.

see a similar field distribution to that seen for the transverse electric component. This is quite a beautiful development. When considering the electric components, they appear to emanate from point sources, the tips and center being the points. When considering the magnetic component, we must look where charge can be in motion, and that can only be in an extended region and not in a point region. Hence, the magnetic field component appears to emanate from the arms of the dipole.

Figure 6.12 shows the wave impedance of the extended dipole, in the space surrounding the dipole.

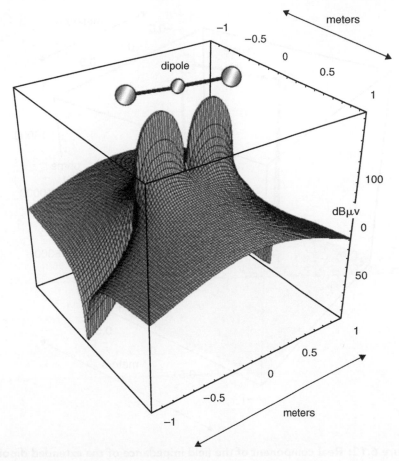

Figure 6.11: Transverse magnetic field component of the extended dipole.

So, now we know why there is no radiation off the dipole tips. It is because the field impedance is heading toward infinity. Through use of the descriptive simple dipole equations, we can build an intuitive understanding of how complex structures radiate. Through observation of impedance fields, we can better understand the directivity of radiation that complex structures will set up. We will cover this topic later in the chapter.

In Figure 6.13, we zoom in more closely as we approach the surface of the dipole radiator. Notice that the real component of the impedance goes to zero as we approach the dipole.

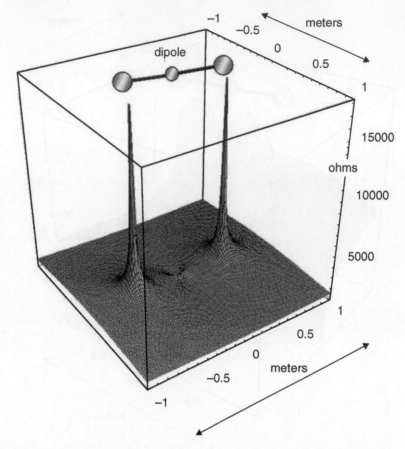

Figure 6.12: Real component of the field impedance of the extended dipole.

As we move away from the dipole, the impedance value approaches 377 ohms, the far-field condition.

Throughout these analytical studies, we've been looking mostly at the real component of the fields. The reason for this is that the real component is the component that will be measured.

We can see that the dipole has a high impedance at the tips; it just does not radiate effectively off the tips. The center impedance of the real component is seen to go to zero, as it should; as we approach the surface of a metal, the impedance should tend to zero. As we move further from the dipole, the impedance should tend to 377 ohms, the far-field

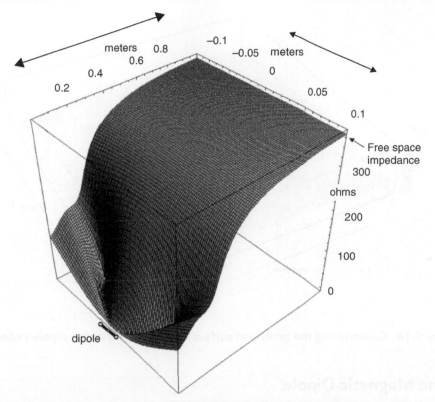

meters   0.8   −0.1

0.6   −0.05   meters

0.4   0

0.2

0.05

0.1

Free space
impedance

300

ohms

200

100

0

dipole

Figure 6.13: Real component of the field impedance of the extended dipole in close to a device.

impedance. Our goal here is to encourage the reader to explore the near field of complex devices. Most textbooks describe the far field, and if you're engaged in developing long-range communication equipment, then you should be familiar with the far field. If you're working in EMC, EMI, RFI, and the mitigation of platform interference effects in wireless systems, then you should be familiar with the near field.

In Figure 6.14, we now show the effect in the near field of many dipole radiators with random phases and orientations. The point of showing this plot is to underline the power and utility of the simple methods outlined here. Beginning with the electric dipole equations, we can write programs to visualize the effects and the field distributions. We show an example of just such a set of analyses performed on measurements of a platform system clock device.

Figure 6.14: Constructing the analytical surface for a set of extended dipole radiators.

6.5 The Magnetic Dipole

We will now move on to a discussion of the magnetic dipole and the importance of the exact placement of probes for measurement of fields generated by devices. Figure 6.15 shows a diagram indicating the orientation of a magnetic dipole with respect to the three Cartesian axes.

For the magnetic dipole, the dipole moment is

$$m_m = \vec{I}\pi b^2 \vec{a}_z, \tag{6.7}$$

where b is the radius of the current loop.

With this, we then obtain that the magnetic vector potential is

$$\vec{A} = \frac{\mu_0}{4\pi} \oint_c \vec{I}\, \frac{e^{-j\beta_0 r'}}{r'} dl$$

$$\vec{A} = \frac{\mu_0 m_m}{4\pi r^2} (1 + j\beta_0 r) e^{-j\beta_0 r} \sin\theta \vec{a}_\varphi. \tag{6.8}$$

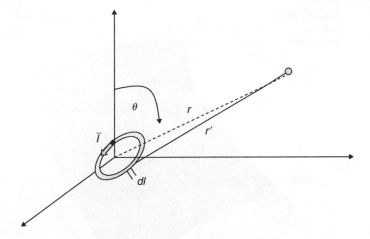

Figure 6.15: The magnetic dipole.

First taking the curl of vector A to obtain the H field, and then taking the curl of the H field, we obtain the following equations (again, we obtain these after some algebraic manipulation, following Clayton Paul's model).

$$\vec{H}_r = j \frac{\omega \mu_0 m_m \beta_0^2 \cos\theta}{2\pi\eta_0} \left(\frac{1}{\beta_0^2 r^2} - j\frac{1}{\beta_0^3 r^3} \right) e^{-j\beta_0 r} \qquad (6.9)$$

$$\vec{H}_\theta = j \frac{\omega \mu_0 m_m \beta_0^2 \sin\theta}{4\pi\eta_0} \left(j\frac{1}{\beta_0 r} + \frac{1}{\beta_0^2 r^2} - j\frac{1}{\beta_0^3 r^3} \right) e^{-j\beta_0 r} \qquad (6.10)$$

$$\vec{E}_\varphi = -j \frac{\omega \mu_0 m_m \beta_0^2}{4\pi} \sin\theta \left(j\frac{1}{\beta_0 r} + \frac{1}{\beta_0^2 r^2} \right) e^{-j\beta_0 r} \qquad (6.11)$$

So, m_m is the magnetic dipole moment.

Also,

$$\vec{H}_\varphi = 0, \ \vec{E}_r = 0, \ \vec{E}_\theta = 0. \qquad (6.12)$$

Note the symmetry between the electric dipole equations and the magnetic dipole equations.

We are interested in the magnetic field that will be measured at a distance r from a device. The radiated component is shown in Figure 6.16, which illustrates the calculated magnetic

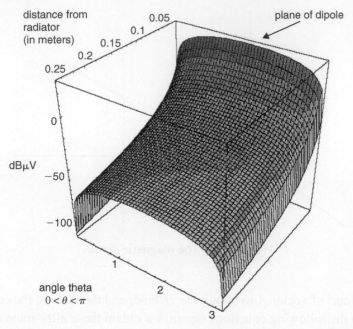

Figure 6.16: Far field component of the magnetic dipole as a function of distance from the radiator.

field as a function of distance, here varying from 1 cm to 25 cm, and as a function of angle deviation from the normal to the device, $0 \rightarrow \pi$.

Figure 6.17 shows the reactive component, $\vec{H}_r$.

The wave impedance, at a particular frequency, is again given as the ratio of the electric field and the magnetic field. For the magnetic dipole, this becomes

$$Z_0 = \eta_{\text{near field}} = \sqrt{\left(\frac{E_\varphi}{H_r}\right)^2 + \left(\frac{E_\varphi}{H_\theta}\right)^2}. \tag{6.13}$$

Figure 6.18 shows the real part of magnetic wave impedance for Equation 6.12.

As the plot shows, in the far field the impedance approaches 377 ohms at $\theta = 0$ and $\theta = \pi$ and falls off to zero as theta varies from this value. Figure 6.19 shows the imaginary portion of the impedance. Note the asymmetry between magnetic and electric dipoles along the

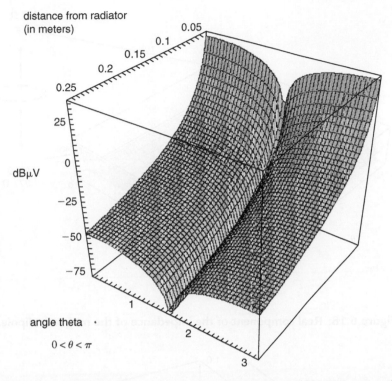

distance from radiator
(in meters)

dBμV

angle theta

$0 < \theta < \pi$

Figure 6.17: Near-field component of the magnetic dipole. Calculated $\vec{H}_r$.

center axis, the electric dipole limit is 377 ohms; the magnetic dipole limit goes to infinity.

Comparison of the impedance distributions for electric and magnetic dipoles indicates that we may be able to infer the type of radiator interior to a larger radiation pattern.

At the risk of being just a touch controversial, sometimes the simulators we use lead us to expect that miracles will occur when we're trying to build models of the systems that we seek to understand. It is the nature of the beast that at some number of points in model development we will find it necessary to make simplifying assumptions, and sometimes these assumptions will be huge. Consider running a simulation of a complex modern motherboard in order to gain some understanding of the radiation mechanisms. The board can easily have several dozen complex ICs, hundreds and possibly thousands of individual circuit nodes, thousands of individual passive devices (with anywhere from 2 to 12 layers in the board construction with complex power geometries on each layer), complex

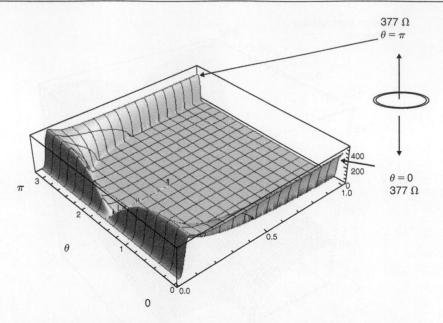

Figure 6.18: Real component of the impedance of the magnetic dipole.

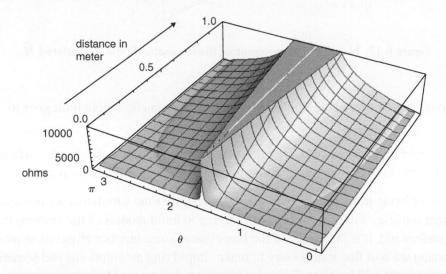

Figure 6.19: Absolute value of the imaginary component of the impedance.

single-ended high-speed traces, and high-speed differential traces. And this just describes the board and not the complex enclosure that the board sits in. Does anyone seriously think that any simulator on the market today, no matter what the cost, is actually performing this analysis in a rigorous manner? If you do, there's a bridge I would like to sell you that spans some wonderful swampland.

So, that said, the models that we're going to discuss now are not meant to be rigorous. They're meant to get us reasonably close to allow us to do some quick what-if analyses, to allow us to gain a quick insight into possible problems, and to provide us some guidance in determining where we should put our limited resources.

6.6 Developing Simple Analytical Structures for Real Devices

Figure 6.20 shows two near-field scans over the silicon die of a system clock device. The scans were performed at 1 GHz and 3 GHz using the electric field probe. This is the same type of device as was shown earlier in the chapter where detailed package field distributions were illustrated.

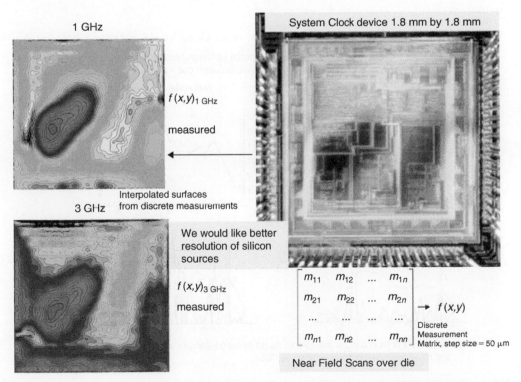

Figure 6.20: Near-field scans of a system clock device: the first step in constructing analytical surfaces.

The plots shown on the left in Figure 6.20 have been image processed to remove any high-frequency spatial noise. The plot shown on the right is the direct near-field scan of the device. It shows quite a bit of activity just southwest of the center. These areas line up with several locations for internal Phase Locked Loops (PLLs) in the device. The matrix at the bottom of the figure is a representation of the discrete set of surface measurements made with the INFS. If we use just the surface scan as is, we still have quite a bit of useful information not available previously. However, if we apply a few analytical surface operators to this set of data, we can derive even better resolution of the areas and regions where we see a high level of emissions.

We will use the Laplacian operator the most often. The Divergence operator can, in some instances, give us useful information related to emissions from the power delivery network.

Figures 6.21 to 6.27 show the development of the analytical method. As mentioned, in some cases the surface measurements can have high frequency spatial variation. This noise

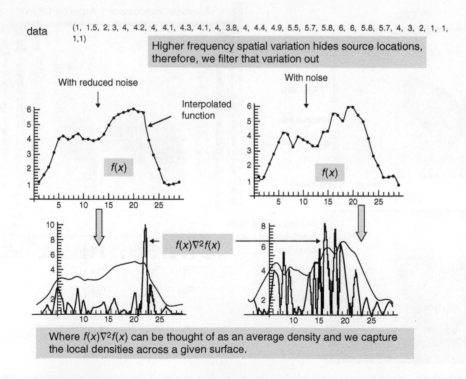

Figure 6.21: High-frequency spatial variation can mask interesting structural details.

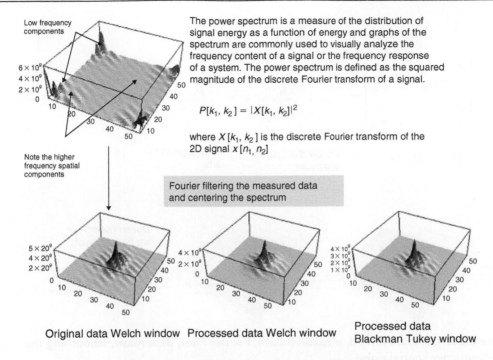

The power spectrum is a measure of the distribution of signal energy as a function of energy and graphs of the spectrum are commonly used to visually analyze the frequency content of a signal or the frequency response of a system. The power spectrum is defined as the squared magnitude of the discrete Fourier transform of a signal.

$$P[k_1, k_2] = |X[k_1, k_2]|^2$$

where $X[k_1, k_2]$ is the discrete Fourier transform of the 2D signal $x[n_1, n_2]$

Fourier filtering the measured data and centering the spectrum

Original data Welch window   Processed data Welch window   Processed data Blackman Tukey window

Figure 6.22: Application of Fourier filtering.

can mask details associated with the source radiators when we use an operator like the Laplacian, which is very sensitive to high-frequency variation. The Laplacian's value is precise because it can amplify those variations associated with underlying structure.

So, the first thing we need to do is remove the extraneous spatial noise. Figure 6.21 shows this where we indicate how the uninteresting spatial variation can mask the structural variation that we do care about.

The general idea is to apply a Fourier transform to the data, center the data in a square matrix, and then apply a filtering mask to remove a prescribed amount of the higher-frequency components of the data set. Figure 6.22 shows the idea and the before and after surfaces. Figure 6.23 shows the smoothing of high-frequency spatial noise.

Note in Figures 6.24 and 6.25 how surfaces that do not vary significantly in their surface details can produce significantly different surfaces after application of the

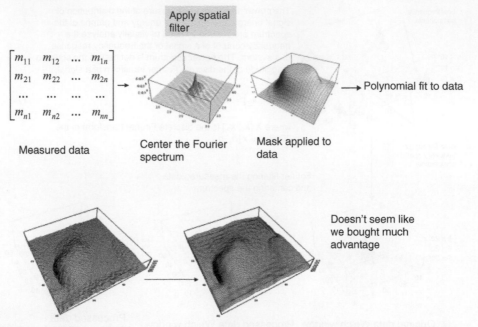

Figure 6.23: General method, measured data, Fourier filter, data mask, smoothing the high-frequency spatial noise.

Laplacian operator. Note also how localized details emerge from the original surface field scan.

In Figure 6.26, note the divergence surface plot in the lower-left corner and how it appears to be divided into two significant lobes. This surface result happens quite often when using the divergence operator. Overall, it appears that we have a dipole radiator. This result can give us some insight into one aspect of the radiation mechanisms present in the silicon. But also note that the divergence operator has highlighted the power delivery rails in the silicon. We're seeing which portions of the power delivery network are radiating the most.

So, we can now apply the analytical operators and build up an understanding of the radiation geometry of the silicon. Figure 6.27 shows how we do just that. Note how we apply different values of the spatial mask filter. With each value of the mask, we add in more spatial variation. At the lowest levels, we see the predominant features. As we increase the mask value, we see higher-frequency spatial features.

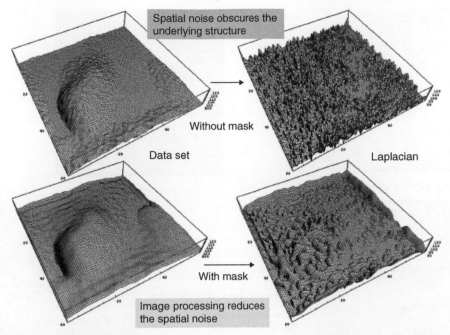

Figure 6.24: Showing how low-level spatial noise can obscure structural details after application of the Laplacian operator.

At the end of this process, it comes down to developing a sense for what value of the mask produces the best surface for interpretation. In Figure 6.27, we see a comparison of two types of surfaces. One is derived from the actual surface measurements, and the other is derived from a complex set of dipole radiators. We wish to develop an approach to complex surface operators that allows us to construct possible radiation surfaces so we can understand the possible interference effects well ahead of device design and application. Comparing the two analytical surfaces in the figure allows us to see that we can build a complex surface from simple dipole radiators that approximates the actual surface as measured when we look at the radiation from a complex silicon surface.

Proceeding along these lines, we define a finite set of point sources that can have a random phase orientation with respect to each other and that can be placed anywhere within the defined area of the virtual device. The idea is shown in Figure 6.28.

We will assume that the radiation sources lie at some distance below the measurement/ evaluation plane. We want to investigate how we can best place point sources in order to

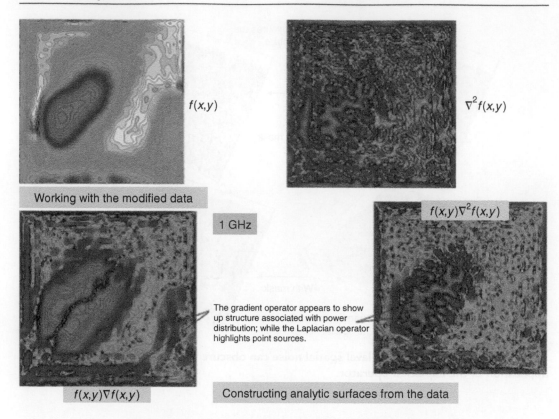

$f(x,y)$

$\nabla^2 f(x,y)$

Working with the modified data

1 GHz

$f(x,y)\nabla^2 f(x,y)$

The gradient operator appears to show
up structure associated with power
distribution; while the Laplacian operator
highlights point sources.

$f(x,y)\nabla f(x,y)$

Constructing analytic surfaces from the data

Figure 6.25: Divergence and Laplacian-derived surfaces.

reduce the radiation surface as measured at the observation plane. This radiation surface
will correlate with far-field emissions as measured in a system implementation. We
indicated in Figure 6.23 that we use a filter to remove higher-frequency features from the
resulting surfaces. In Figures 6.29 and 6.30, we show the effect of two different mask
values. In the lowest mask value, we see that the exact location of the point sources
is smeared out as compared to when we increase the mask value and include
higher-frequency surface features. The greater the mask value, the more exactly we can
locate the radiation surface features. However, when we investigate radiation surfaces
using different mask values, we will see different structures come into play. And that's
what we want. Certain values of the mask will show us the power distribution structure as
it relates to radiated emissions, and other values of the mask will reveal radiated emissions
directly related to point sources, such as would be associated with functional areas such as
PLLs.

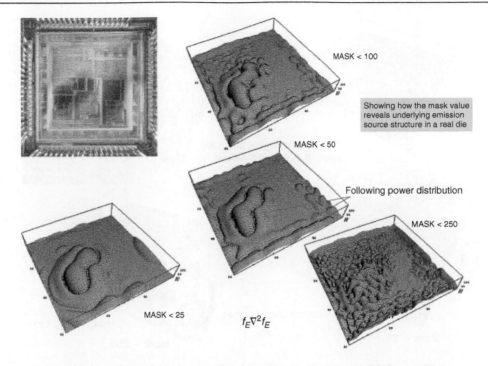

MASK < 100

Showing how the mask value reveals underlying emission source structure in a real die

MASK < 50

Following power distribution

MASK < 250

MASK < 25

$f_E \nabla^2 f_E$

Figure 6.26: Varying the mask value and the resulting analytical surfaces.

So, we'll develop this idea. We will now start analyzing finite sets of point sources. But we will add an additional analytical option. We will use the idea of the Lagrangian lifted from classical mechanics and from quantum electrodynamics. We will not spend a great deal of time justifying this approach. We will reference previous work in this field for the interested reader, and proceed to try the method and then compare measurement against theory.

Classical mechanics operates with two types of parameters; a complete system of equations of motion and a complete set of initial conditions. The *Lagrangian* is defined as the sum of kinetic and potential energies.

$$L = KE - PE = T - V$$

The kinetic energy, T, is a function of $\dot{x}$, but not of x, whereas the potential energy, V, is a function of x, but not of $\dot{x}$.

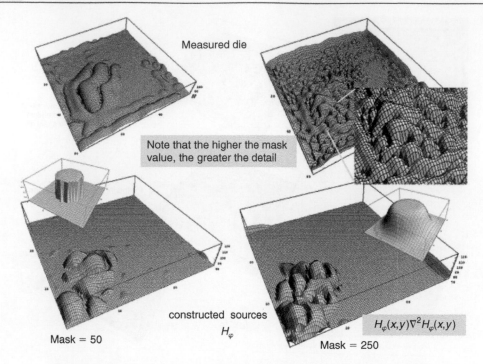

Measured die

Note that the higher the mask value, the greater the detail

constructed sources
H_φ

$H_\varphi(x,y)\nabla^2 H_\varphi(x,y)$

Mask = 50

Mask = 250

Figure 6.27: Comparing surface details as a function of mask size.

The Lagrangian has the property that

$$\frac{\partial L}{\partial \dot{x}} = \frac{\partial T}{\partial \dot{x}}$$

$$\frac{\partial L}{\partial x} = -\frac{\partial V}{\partial x}. \tag{6.14}$$

Proceeding by analogy, since the magnetic field, $\vec{H}$, is proportional to $\dfrac{dq}{dt}$, and $\vec{E}$ to q, we have for the Lagrangian

$$L\left(\dot{q}, q\right) = T\left(\dot{q}\right) - V\left(q\right). \tag{6.15}$$

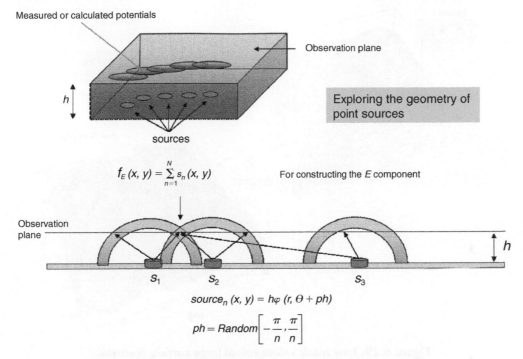

Measured or calculated potentials

Observation plane

Exploring the geometry of point sources

h

sources

$$f_E(x, y) = \sum_{n=1}^{N} s_n(x, y)$$

For constructing the E component

Observation plane

S_1 S_2 S_3

h

$$source_n(x, y) = h\varphi(r, \Theta + ph)$$

$$ph = Random\left[-\frac{\pi}{n}, \frac{\pi}{n}\right]$$

Figure 6.28: Composing a surface of multiple point sources.

We then assign the mean normalized arrays of the measured field quantities to the kinetic and potential energy terms

$$L(\dot{q}, q) = \langle |H| \rangle^2 - \langle |E| \rangle^2 \tag{6.16}$$

$$= H_\varphi(x, y)\nabla^2 H_\varphi(x, y) - E_\varphi(x, y)\nabla^2 E_\varphi(x, y),$$

where $\langle |H| \rangle$ is the summation of each measured array of the magnetic field, mean normalized, at a discrete set of frequencies, and similarly for the electric field array. In constructing the summation, we have assumed a weighting factor equal to 1.

After forming the Lagrangian, we integrate over the region of interest, which in our work will be the die and package of a system device. In the analysis given here, we are placing a set of point sources in a region approximating the size of a clock die. The integral is called the *action integral* and for our purposes is a measure of the potential for radiated

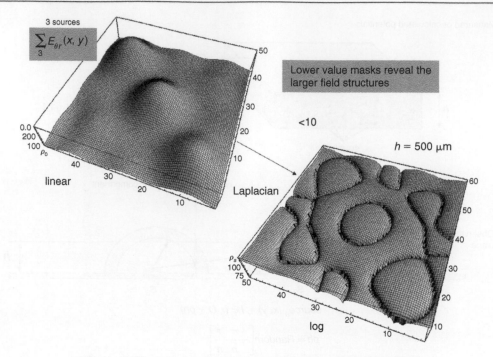

Figure 6.29: Low mask values reveal large surface features.

emissions. Therefore, the lower the action integral value the better. This is shown in Figure 6.31.

Figure 6.32 shows a comparison of two sets of source points; the first distributes the point sources over the entire area, and the second concentrates the point sources in a small region of the area under study. As shown, the action integral difference between the two conditions is quite dramatic. It also shows that when the sources are concentrated close to each other, the expected radiation surface will be significantly higher than it would be for distributed sources.

In Figure 6.33, we show a comparison of the analytical and the measured results.

So what's the point? We want to have a simple, relatively speaking, method of analysis. Anyone who has the money can buy an expensive 3D field simulator and then try to feed into it all of the geometry of their problem—all the nodes, all the traces, all the layers of signals and power. But nowhere has anyone ever shown that the simulators will converge to a single true answer if given a sufficiently complex problem like a typical PCB.

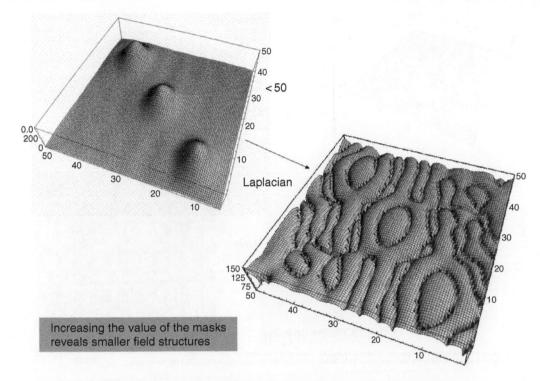

Figure 6.30: Larger mask values reveal finer surface structure.

What we've been building here is an approach that does not depend upon tens of thousands of dollars, nor does it assume a complete answer exists for these complex problems that we wish to solve. We have shown that you can get close, within 90% or so, by using your head and trying to pull out the essential features of the problem. If you do this, you can get close. You have to ask yourself, do you really need 100% accuracy?

Figure 6.34 shows how we can use the method developed to this point to begin to lay out design guidelines for silicon and packages. We start with a point source distribution that closely represents a set of point sources as measured in a real device. We then start moving the sources apart from each other in order to reduce the radiation surface potential. Through such means, by only moving the sources (not reducing the intensity of the sources or reducing the number of the sources), we can gain a 2 dB to 3 dB decrease in the potential for radiated emissions. While that may not seem dramatic, we must always remember that these are usually additive, and every little bit helps.

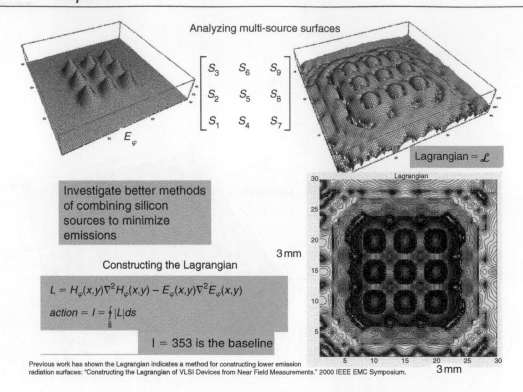

Figure 6.31: Constructing the lagrangian of point sources.

In the next example, we use the model to see what kind of changes occur when we can impose random phase conditions between the point sources. In Figure 6.35, we can see that the random phase condition compared to the coherent phase can produce a radiated emissions difference of up to 44 dB—and that's huge.

6.7 Developing Channel Models

The following example of model building uses PCIe. In Figure 6.36, we've built a simple differential model of a PRBS signal using a system simulator SystemView. We show two instances: with and without skew between the differential pair. As shown, even as little as 10 pS of skew can lead to a difference of 14 dB at the data rate of 2.5 Gb/s. The data rate is where the signal null occurs. Most people might think that a signal null means just that; there isn't any signal energy there. Alas, this is not the case. If you could guarantee a signal with absolute pure symmetry, then you could have a signal with a deep null with no energy

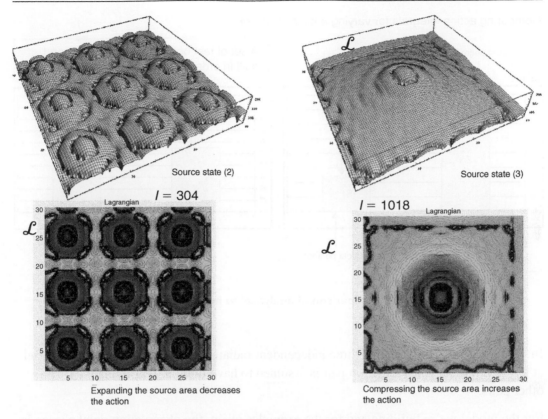

Source state (2)

$l = 304$

Source state (3)

$l = 1018$

Expanding the source area decreases
the action

Compressing the source area increases
the action

Figure 6.32: Comparison of two sets of source points.

present. But you can't, and as we saw in Chapter 2, any signal asymmetry leads to signal energy in the null. In fact, what we're seeing is a conversion of differential energy to common mode energy as we approach the signal nulls. The simulator is a good starting point and allows us a means to check the analytical models that we may build.

In order to build our analytical model, we will zoom in on the PRBS signal. The next few figures indicate the idea.

In PCIe, a single data event—the unit interval (UI)—is 400 pS in duration. This is shown in Figure 6.37. Also indicated is the skew that is present between the transmit plus, D+, and the transmit minus, D−. For the model we develop we'll assume that the channel length is 0.3 m. Typical transition time on such a channel is 1 nS. We will further assume that each edge transition is a radiation event.

Comparing action integrals for varying source structure

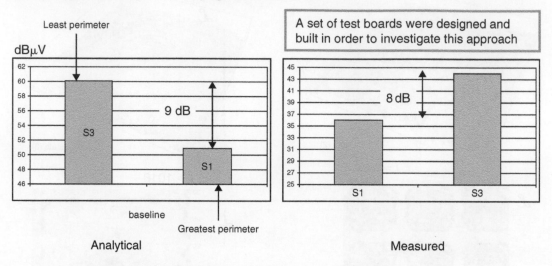

Figure 6.33: Comparison of analytical to measurement results.

In Figure 6.38, we assume that three independent radiating pairs can exist on the channel at any given time. Each radiation pair is assumed to have random phase relative to the other pairs.

For the model we're building and for the examples given, the observation point chosen is 6 inches above the plane containing the radiation channel. Throughout our discussion, we will assume that the excitation sources are sinusoidal. This is shown in Figures 6.39 and 6.40.

The description of the excitation source is shown in Figure 6.41, where we incorporate the frequency, the phase of the channel radiators relative to each other, and the phase relation between the channels.

A point to be emphasized here is that the model so far is not that complex. Rather, it is quite simple. It assumes a set of differential radiators, with sinusoidal excitation, with variable phase between the radiators and the channels. It further assumes that the channel geometry is straight. So, the model is simple and quick. The question to ask is whether it actually produces answers that are useful. We'll show some results shortly. First, though, we will finish putting the model together and show some calculational results and comparisons. The model is shown in Figure 6.42.

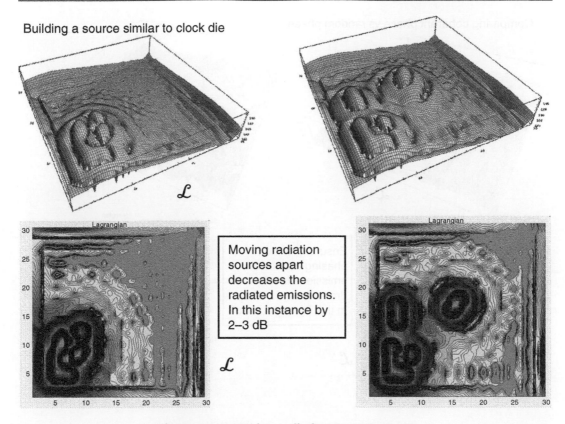

Figure 6.34: Moving radiation sources apart.

The model as developed was analyzed using Mathematica. Since we have assumed random phase components, we run the model for over 2,000 trials and determine the distribution of radiated emissions that would be seen at the observation point. We show two instances: a single radiator and three radiators. We provide a random skew between the pair edges and between the channels in addition to the random phase of the sinusoidal excitations. As shown in Figure 6.43, there is a 6 dB difference in the mean emissions level between a single radiation pair and a set of three radiation pairs.

Figure 6.44 extends the analytical results of the model to multiple channels and increased skew. We can see that the mean emissions levels increase as the number of channels increases, which makes intuitive sense. However, due to the random nature of their

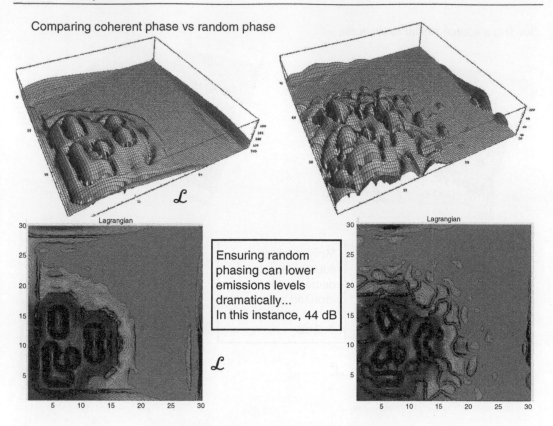

Figure 6.35: Random phase between sources.

combination, the increase is not linear, but rather goes as the square root of the number of the channels. Also note that as the number of channels increases, the distribution approximates a Gaussian more closely.

What does this approach give us that just using a numerical solver doesn't? Model building in this manner allows us to explore those variables that we think are important to control and allows us to construct models with a set of variables that we wish to vary and develop an understanding of their dependence. Recall from Chapter 3 that we did the same type of Monte Carlo analysis for signal structure, looking at the RFI impact when we considered edge rate jitter and pulse width jitter.

We've seen that skew can be an important factor in platform interference. We've seen that simple time displacement between two halves of a differential channel will create the skew

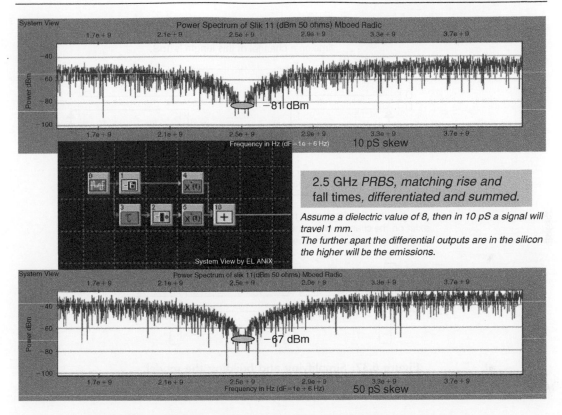

Figure 6.36: Building a differential model of a high-speed data stream.

that generates radiated emissions. But time displacement is not the only generating means; physical displacement between the two transmitters of the differential signal can also create skew. We show the idea in Figures 6.45 to 6.47.

In these examples, we assume a random time skew of 45 to 55 pS. We also assume that the victim circuit is in very close proximity, in these instances being coplanar and sharing the same substrate. We show the difference in the resulting radiating emissions when the transmitter pair is separated by 1 mm and by 5 mm. The resulting emissions differ by up to 17 dB, a rather significant increase. Now, 1 to 5 mm may be quite a large distance of separation when considered in silicon, but it may not be that significant when the transmission pairs are composed of discrete parts. Even in silicon, we want to consider component placement and their relative separations. Differential skew is a parameter that we can have direct control over and we should exercise that control as positively and as

Consider a single PCIe diiferential channel, D+ and D−, with a length, $l = 0.3$ m, UI = 400 pS. Typical transmission time on a channel of this length is 1nS.

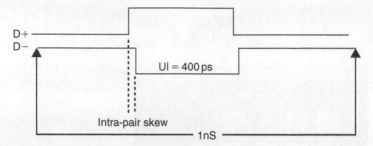

Assume that 3 independent point source radiator pairs exist on the channel with random phase relative to each other, and constant channel skew.

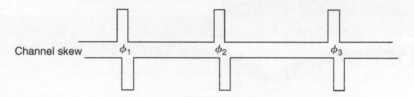

Figure 6.37: Simplifying and building the model.

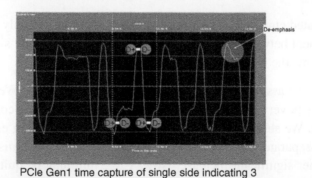

PCIe Gen1 time capture of single side indicating 3 independent radiating pairs

Figure 6.38: Breaking the signal structure down to independent radiators.

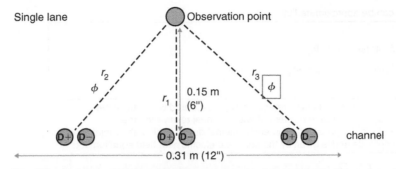

Single lane

Observation point

r_2
ϕ

r_1

0.15 m
(6")

r_3
ϕ

D+ D−   D+ D−   D+ D−   channel

──────── 0.31 m (12") ────────

The difference in propagation time between r_1 and r_2 (and r_3) is 53 pS

Figure 6.39: Further development of the channel radiation model.

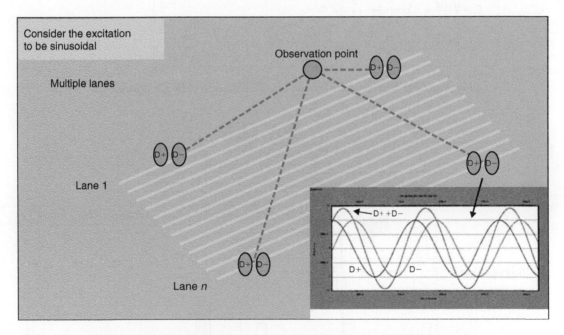

Consider the excitation
to be sinusoidal

Multiple lanes

Observation point

D+ D−

D+ D−

D+ D−

Lane 1

D+ D−

Lane *n*

D+ +D−

D+   D−

Figure 6.40: Extending the model to encompass multiple channels.

often as possible. Again note that as the number of channels increases, the emissions measured at an observation point also increase.

We will now build a channel model using a commercial system simulator. This particular simulator is available from Agilent and is SystemView. We can either use an analytical

Each pair can be approximated by

$$D+ = E_0 \sin(\omega_0 + \xi + \Phi_R)$$
$$D- = E_0 \sin(\omega_0 + \sigma + \xi + \Phi_R + \pi) \qquad (1)$$

Where σ is the phase shift of D− relative to D+ associated with transmitter
output offsets, Φ_R is a random phase of each channel relative to other
channels and ξ is the phase offset in each channel due to each of the radiator pairs.
E_0 is the amplitude and is given by the complete electric near field equation set:

$$E_0 = \sqrt{E_\theta^2 + E_r^2} \text{ where}$$

$$E_\theta = \frac{\hat{I}dl}{4\pi} \eta_0 \beta_0^2 \sin\left(\frac{j}{\beta_0 r} - \frac{1}{\beta_0^2 r^2} + \frac{j}{\beta_0^3 r^3}\right) e^{-j\beta_0 \bar{r}}$$

$$E_r = 2\frac{\hat{I}dl}{4\pi} \eta_0 \beta_0^2 \cos\left(\frac{1}{\beta_0^2 r^2} + \frac{j}{\beta_0^3 r^3}\right) e^{-j\beta_0 \bar{r}}$$

$$\eta_0 = 377\,\Omega \quad \beta_0 = 2\pi\Big/\lambda \qquad (2)$$

Hertzian electric dipole
equations

Figure 6.41: Building the model using elementary dipole equations.

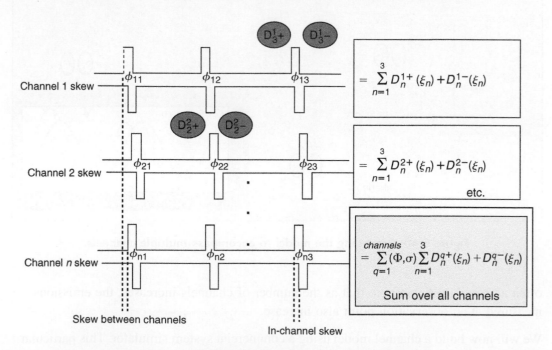

Figure 6.42: Putting the model pieces together.

Analysis for E_θ only.
V derived from measurements with 100 KHz RBW.
For 1 channel with 1 and 3 independent radiator pairs the following plots show the distributions.

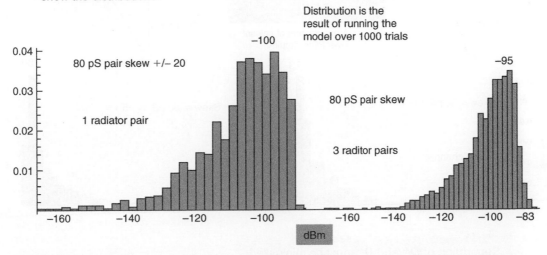

Figure 6.43: Comparison of radiated emissions between single and multiple pairs of sources.

Channel skew will increase the emissions in multi-channel analysis

Let the channel skew = 200 pS +/− 40

For 16 channels with 3 independent radiator pairs

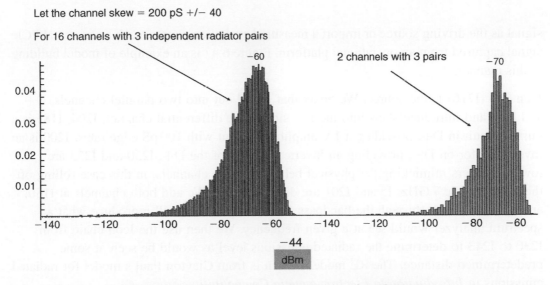

Figure 6.44: Multiple channels of source interference.

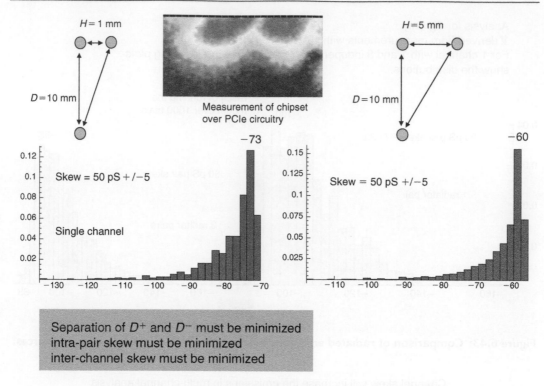

Separation of $D+$ and $D-$ must be minimized
intra-pair skew must be minimized
inter-channel skew must be minimized

Figure 6.45: Equating skew with physical channel separation.

signal as the driving source or import a measured excitation source, such as an actual PCIe signal captured from an operational platform. Figure 6.47 is an example of model building in this manner.

Icon 216 (I216) is the source. We break that signal out into two parallel channels, D+ and D−, and introduce skew into the D− side of the differential channel, I204. I185 is a buffer circuit in D+, providing a 1 V amplitude output with 100 pS edge rates. I200 is an inverse buffer on D−, providing an inverted version of the D+. I250 and I253 are low-pass filters mimicking the physical behavior of the channels, in this case rolling off the signal above 7 GHz. I5 and I201 are differentiators. We add both channels at I180, and send the result through the bandpass filter at I10, which will reproduce what a spectrum analyzer would see at a given frequency. We then use the Icon chain from I230 to I243 to determine the radiated emissions level as would be seen at some predetermined distance. The RE model chosen is from Clayton Paul's model for radiated emissions in *Introduction to Electromagnetic Compatibility*.

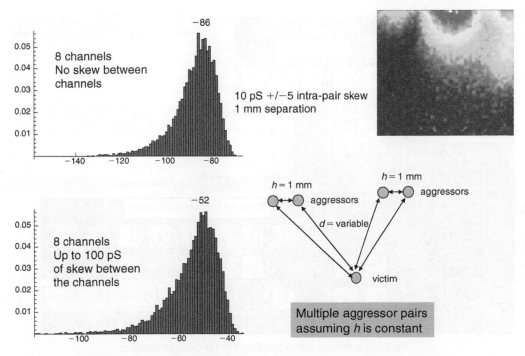

Figure 6.46: Multiple channels with constant physical separation.

This is, again, a relatively simple model to create. You should probably ask how well it compares against real emissions seen from a real test structure. Figures 6.48 and 6.49 show a set of comparisons of model versus measurement. The GTEM cell was used for this comparison. The test structure was a single differential channel driven by a high-speed pattern generator that can add skew between the D+ and D− sides. Skew can be controlled to within 1 pS to 2 pS. The resulting radiated emissions were measured in the GTEM and are shown in the figures.

As we can see, in this instance model and measurement are in almost exact agreement. This is not usually the case, but it is gratifying to see that we can achieve quite good predictive abilities from rather simple analytical approaches.

Figure 6.49 zooms in on the first few hundred of picoseconds skew (the highlighted region in Figure 6.48) to provide a closer examination.

Note that this comparison is looking at the fifth harmonic of the excitation source and that there is only one channel being measured, and hence no channel skew to account for. We

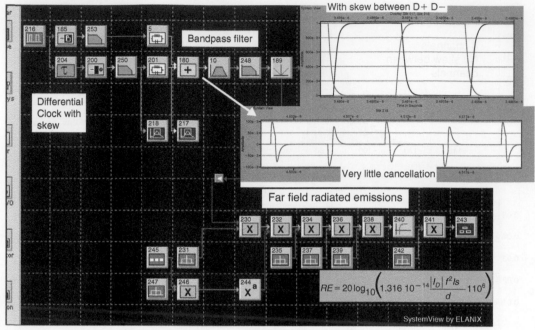

Building a model for predicting the far-field emissions
from a differential clock with skew between the pair.

Figure 6.47: A radiated emissions model using Agilent's SystemView.

can use these models and extend them to look at multiple channels. Figure 6.50 is an analytical comparison of the effect of channel skew and harmonic number. In this case, we construct a surface that integrates the output of I241 from Figure 6.47. As we see, harmonic value and channel skew can have a considerable impact on the resulting emissions. Bear in mind at this point that we might be able to take advantage of these facets and design accordingly. For example, we might note that channel skews of 400 pS and 1,000 pS produce the lowest overall emission levels through the harmonic range of interest. Also note that there are harmonic ranges where channel skew has little impact at all. Recall from Chapters 2 and 3 that signal structure could be used to reduce the RFI of the platform. By combining signal structure experimentation with platform channel variation, we can build complex models that track the RFI envelope quite well.

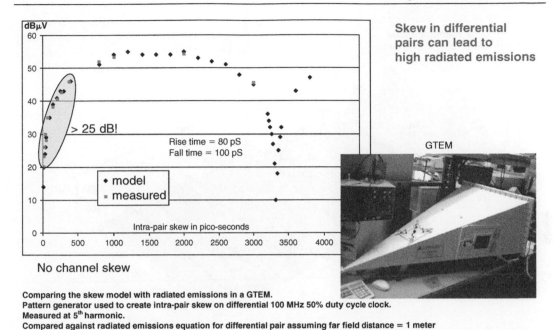

Skew in differential
pairs can lead to
high radiated emissions

GTEM

No channel skew

Comparing the skew model with radiated emissions in a GTEM.
Pattern generator used to create intra-pair skew on differential 100 MHz 50% duty cycle clock.
Measured at 5th harmonic.
Compared against radiated emissions equation for differential pair assuming far field distance = 1 meter

Figure 6.48: Comparing model versus measurement.

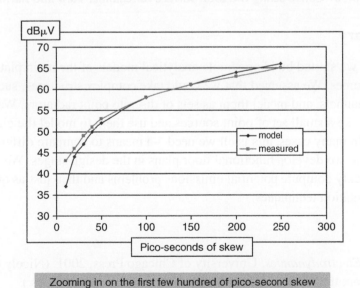

Zooming in on the first few hundred of pico-second skew

Figure 6.49: A closer examination of model versus measurement.

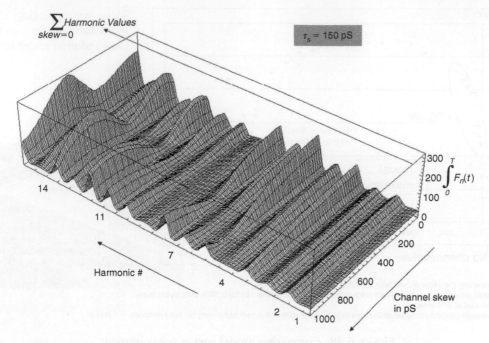

Figure 6.50: Constructing the total surface for channel skew and harmonics.

6.8 Summary

In this chapter, we started building simple analytical models of the main platform interference sources. We assumed that we could take complex structures, such as the interconnect channels, and model them as sets of discrete point radiators. We also showed how we could take a small set of point sources and use them to model the circuit sources inside silicon. In many cases, this is all we need—a means to compare different source configurations as we develop functional floor plans in the design stages. We can use these methods to quickly compute potential emissions problems and then just as quickly investigate mitigation techniques.

References

[1] F. Melia, *Electrodynamics*, University of Chicago Press, 2001. (Nicely done presentation; the physics is developed along with the mathematics.)

[2] R.P. Feynman, R.B. Leighton, and M. Sands, *Feynman Lectures on Physics*, vol. 2, Addison-Wesley, 2006. (A must-have on every engineer's bookshelf.)

[3] W.A. Blanpied, *Modern Physics: An Introduction to Its Mathematical Language*, Holt, Rinehart and Winston, 1971.

[4] J. Vanderlinde, *Classical Electromagnetic Theory*, 2nd ed., Springer, 2004.

[5] R.E. Hummel, *Electronic Properties of Materials*, 2nd ed., Springer-Verlag, 1993.

[6] C. Paul, *Introduction to Electromagnetic Compatibility*, Wiley, 1992.

[7] C. Paul and S. Nasar, *Introduction to Electromagnetic Fields*, McGraw-Hill, 1987. (Both Clayton Paul books are essential references that should be close at hand.)

[8] S. Ben Dhia, M. Ramdani, and E. Sicard, *Electromagnetic Compatibility of Integrated Circuits*, Springer, 2006.

[9] A. Taflove and S.C. Hagtness, *Computational Electrodynamics*, Artech House, 2005.

[10] B. Sklar, *Digital Communications*, Prentice Hall PTR, 2001.

[11] J.D. Jackson, *Classical Electrodynamics*, Wiley, 1999.

[12] K.A. Milton and J. Schwinger, *Electromagnetic Radiation: Variational Methods, Waveguides and Accelerators*, Springer, 2006.

[13] Jerry P. Marion, *Classical Dynamics of Particles and Systems*, Academic Press, 1965.

[14] K. Slattery, J. Muccioli, and T. North, *Constructing the Lagrangian of VLSI Devices from Near Field Measurements of the Electric and Magnetic Fields*, IEEE 2000 EMC Symposium, Washington, DC.

[15] K. Slattery and W. Cui, Measuring *the Electric and Magnetic Near Fields in VLSI Devices*, IEEE 1999 EMC Symposium, Seattle.

[16] K. Slattery, X. Dong, K. Daniel, *Measurement of a Point Source Radiator Using Magnetic and Electric Probes and Application to Silicon Design of Clock Devices*, IEEE 2007 EMC Symposium, Honolulu.

[17] K. Slattery, J. Muccioli, T. North, *Modeling the Radiated Emissions from Microprocessors and Other VLSI Devices*, IEEE 2000 EMC Symposium, Washington, DC.

CHAPTER 7

Connectors, Cables, and Power Planes

7.1 Chance and Caprice Rule the World

A major source of emissions can originate from connectors, cables, and power planes. Of course, it is well known that cables can be significant sources for radiated emissions. However, at the frequencies that we're concerned with in investigating platform interference in wireless systems, the cable connectors themselves can also be significant sources of interference. We will now build a model of a typical high-speed connector for high-definition media interface (HDMI) systems. Figure 7.1 shows a 3D model constructed for use in Ansoft's HFSS. We show surface currents as they would be seen inside the metal shell of the connector. We also show the radiation patterns at two

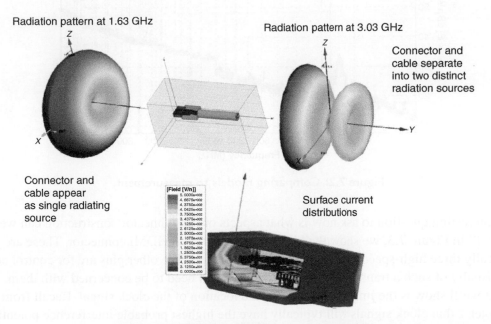

Radiation pattern at 1.63 GHz

Radiation pattern at 3.03 GHz

Connector and cable separate into two distinct radiation sources

Connector and cable appear as single radiating source

[Field [V/n]]

Surface current distributions

Figure 7.1: Building HDMI connector models.

frequencies, which correspond to data rates that would be transmitted over an HDMI link. As we can see, the patterns vary with frequency, with the lower frequency pattern appearing more isotropic than the higher frequency pattern. As the frequency of excitation increases, this is typically the case; the pattern becomes less isotropic and more directive.

Figure 7.2 shows a comparison from 2 MHz to 6 GHz. As indicated, the agreement between the model we built and the measurements is very good. All frequencies show agreement within a few decibels, except at one frequency where the measurement and model diverge by 10 dB. Overall, there is quite good agreement. This level of agreement gives us confidence that our analytical investigations will agree with measurement and will correlate when implemented in a real platform.

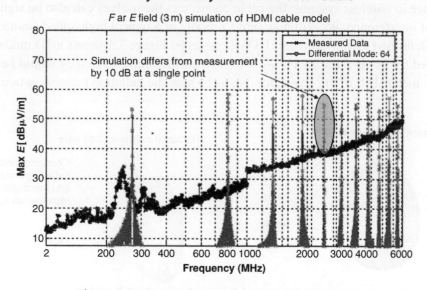

Figure 7.2: Comparing models to measurement.

An interesting question to ask now is what aspects of the connector construction can we impact? In Figure 7.3, we show the pin assignments for an HDMI connector. These are typically three high-speed data pairs and one clock pair. The other pins are for control and are usually of such a transmission speed that we don't need to be concerned with them. What we'll show is the impact of changing the location of the clock signal. Recall from Chapter 2 that clock signals will typically have the highest probable interference potential and the highest EMI potential.

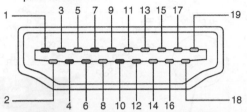

Effects of changing the position of the clock pair

The 4 possible positions for the clock pair are:
- pins 1 and 3
- pins 4 and 6
- pins 7 and 9 and
- pins 10 and 12.

The pin assignment of the HDMI connector showing the position of the 4 differential pairs

Figure 7.3: HDMI connector and high-speed data pair assignments.

As noted, there are four possible positions to place the clock signal, ranging from the approximate center of the connector to the very corner of the connector. So, the possibilities range from near symmetry to quite asymmetric. Which is the best? Intuitively, one would think that symmetry is the way to go.

In Figure 7.4, we show a comparison of the measured radiated emissions from an HDMI connector when a 38 MHz clock is located at each of the four high-speed data pair locations.

As expected, the data pair location furthest from the center of the connector produces the highest emissions levels. The data pair location closest to the center produces the lowest emissions levels.

Figure 7.5 is another measurement of the impact of data pair skew on the radiated emissions levels. The previous skew measurements showed a comparison of model and measurement at a single frequency. This figure shows measurements over a wide band of frequencies and indicates that skew effects hold over a wide range of frequencies.

7.2 Power Distribution Radiated Emissions

Our last example of model building and measurement concerned developing models for power delivery in PCBs. In Figure 7.6, we show two of the possible resonant modes of a power plane construction. In this example, the PCB is a 30 mm by 50 mm mini-PCIe form factor board used to add on wireless functions. It's a small board, and that's part of the

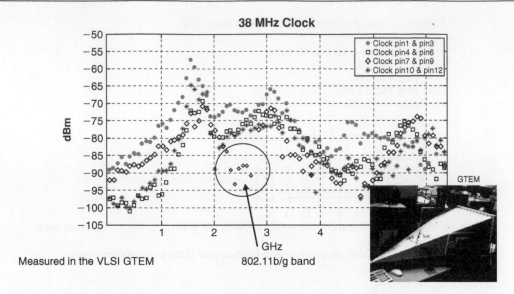

Figure 7.4: Measured radiated emissions due to forwarded clock.

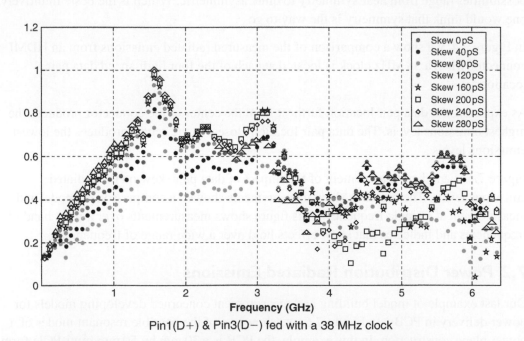

Pin1(D+) & Pin3(D−) fed with a 38 MHz clock

Figure 7.5: Measuring the effect of data pair skew on the radiated emissions.

mPCle card

First resonant mode: $m = 1$, $n = 0$, **Freq = 2.46 GHz**

Higher order mode: $m = 2$, $n = 1$, **Freq = 5.15 GHz**

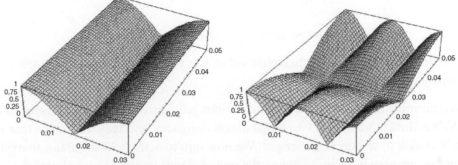

**Analytical solution for power plane resonances
using Mathematica (Source: IEEE EMC Transactions, August 2003)**

Figure 7.6: Resonant modes in power plane structures.

problem. One side of the board has the very high speed interconnect circuitry for PCIe, and the other side has radio circuitry such as mixers, low-noise amplifiers, and other sensitive circuits. At best, physical separation between a very high noise source and a very sensitive to noise receiver is about 25 mm or so. We've seen previously that we need to guarantee certain amounts of system isolation in order to maintain radio system performance. How can we do this?

We can start by analyzing resonance structures in our PCB, determining which wireless frequencies may have problems due to board construction. In the instance given here, we see that the size of the add-in card produces resonances in at least two very important radio bands, 802.11a/b/g.

These are the equations used to calculate the power plane resonances.

$$kx(m) = \frac{m\pi}{\text{length}}, \quad ky(n) = \frac{n\pi}{\text{width}},$$

$$\text{quality}(\omega) = \frac{1}{\delta + \frac{1}{h}\sqrt{\frac{2}{\omega\mu_0\sigma}}}$$

$$\text{resonant}\,V(x, y) = \cos(kx(m)x)\cos(ky(n)y)$$

$$\text{voltage}(x, y) = \frac{120}{\sqrt{\varepsilon_r}}\frac{h}{\text{width}}\frac{2\cos(kx(m)x_0)\cos(ky(n)y_0)}{\sqrt{m^2 + n^2\left(\frac{\text{length}}{\text{width}}\right)^2}}\text{quality}(\omega)$$

$$\text{voltageDist} = \text{resonant}\,V(x, y)\text{voltage}(x, y)$$

where $\delta \equiv$ loss tangent, ε_r is the dielectric value, h is the distance between planes, μ_0 is the permeability and σ is the conductivity.

We first determine that we may have some serious problems with the form factor of the PCB. We've already gotten back to the platform designers, and they've told us that we have to live with it; it won't be changed. We now turn to a 3D field simulator to explore what options we may have in designing the power planes in order to maximize the isolation that good PCB construction can deliver to us.

There are many ways to approach such a study. We may have leeway in the number of signal and power planes that we can use, or we may be restricted to four layers for both signal and power. Let's assume that we've been restricted to only four layers due to cost constraints. We then need to know how we can deliver power to the high-noise digital circuitry and the low-noise analog circuitry. We create the simplest geometry first. We remove all unnecessary features in order to build a model we can believe. When we arrive at a model PCB that we can believe, we can then start adding the smaller features.

What we're interested in is what happens when the digital circuitry excites the structural resonances. So we locate different positions for the excitation and for the receiver and find those locations that give the best results (or the worst results). In the example shown in Figure 7.7, we locate both excitation and receiver along an axis of symmetry.

The power of the simulators lies in being able to show us how power distributes itself on the inner layers. With the INFS, we can measure field distributions at the surface of the PCB, but the INFS gives us nothing on the interior field distributions. If we can show correlation between what we can measure at the surface and what the simulator is showing us, then we can gain confidence about what the simulator is telling us regarding the interior conditions. Figure 7.8 shows the general form of the power plane model that we will use. In Figure 7.9 we see a power distribution at 2.46 GHz as given by the simulator for a four-layer construction with solid power and ground planes. This conforms pretty well with what we expected due to the analytical treatment shown earlier. So, with some confidence, we then look at how the power is flowing on the interior layers.

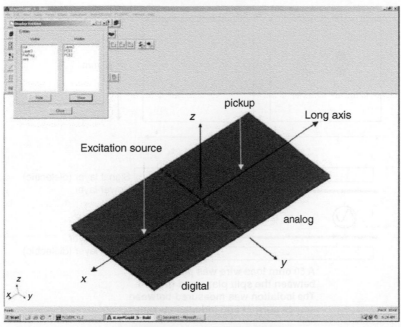

Figure 7.7: Constructing a test PCB for power plane exploration.

Moving on from the solid power and ground planes, we begin to investigate modifications to the power distribution structure. Specifically, we separate the digital and analog portions and provide continuity only at two narrow locations at the top and bottom of the board. Figure 7.12 we show the simulation comparison in Figure 7.10 between the solid planes and the segmented planes. Implementing something as simple as a gap between the power planes for digital and analog sections leads to a 30 dB increase in the achievable isolation. The cost to us is zero. There are no added parts and no added layers—just a different approach to implementing how we route the power planes. In Figure 7.11, we show a wideband comparison of the two approaches to power plane structure. On average, over a 5 GHz range, we get 10 dB of added isolation, and in some cases we see up to 30 dB of added isolation.

We'll now start looking at the impact to radiated emissions potentials when we combine geometries, such as when we add thermal solutions to existing device structures. We will

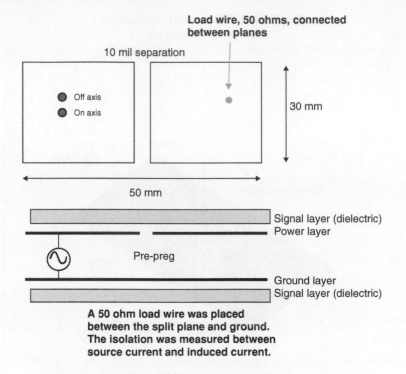

Figure 7.8: Locating digital noise source and analog receiver.

Figure 7.9: Radiated power distributions due to solid power plane resonance.

PCB Models: Experimenting with isolation schemes
Achievable Isolation improvements: 4 layers

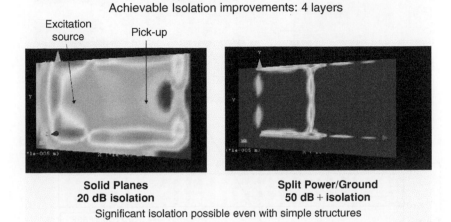

Solid Planes	Split Power/Ground
20 dB isolation	50 dB + isolation

Significant isolation possible even with simple structures

Figure 7.10: Isolation comparison due to power plane construction.

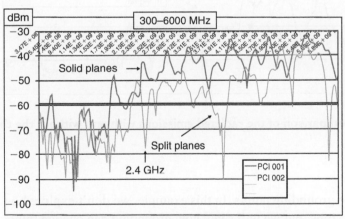

Experimental measurements of PCBs: Solid versus split planes—
separating the digital from the radio power gains 10 dB of isolation
on average and up to +30 dB in some ranges

Figure 7.11: Measurement of test PCBs to validate the 3D simulations.

first show a simple measurement. We'll take one of our test structures, which we will describe shortly, and then add an existing thermal solution found inside a commercial notebook. We can immediately see that there can be 40 dB increases in the emissions when a floating thermal solution is added. We also note that the thermal solution shown here has

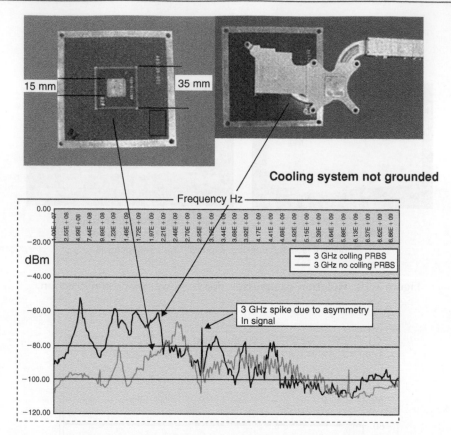

Figure 7.12: A comparison of the radiated emissions with and without the thermal solution.

provision for five grounding points. In this example, we don't explore how grounding impacts the emissions, but we will show how grounding can change things, sometimes not for the better.

We'll continue our study of radiation structure by considering a set of experiments that look at the impact of shape on heat spreaders. Figures 7.13 and 7.14 show the geometry of a typical heat sink and device and indicates the capacitive coupling that exists between the device and the heat spreader and the heat sink. We'll extract the significant geometry to produce something that can be modeled relatively easily. We'll then build both the simulation models and the experimental models. We show the five shapes that we investigated in Figure 7.15.

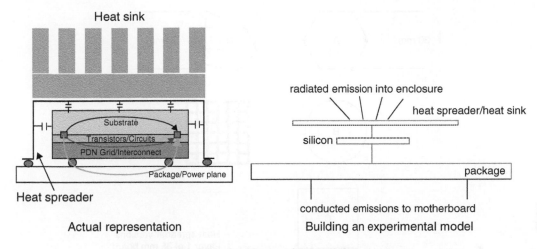

Figure 7.13: The significant pieces of the geometry.

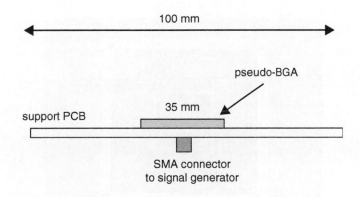

Figure 7.14: Building a test board for GTEM radiated emissions.

The pseudo-BGA is a separate test board, 35 mm square, of four to six layers executed in FR-4. It is a cheap way to mimic a BGA package attached to a PCB. Designing and building a true BGA package easily runs to $50,000. This test structure is more like $1,000. We can therefore make many different test structures for a reasonable amount of money. Also, we need to only make a couple of the larger GTEM test boards and concentrate on the smaller 35 mm test board to make many different configurations.

We chose the shapes in Figure 7.15 because they closely resemble the shapes commonly used in commercial products today. Shapes D and E have been added to investigate how

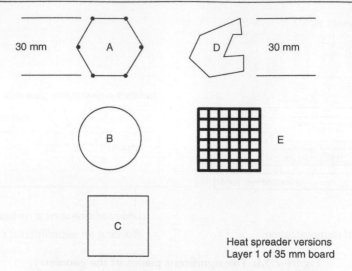

Figure 7.15: Experimental shapes for heat spreaders.

Figure 7.16: The GTEM test board with the pseudo-BGA PCB.

strong deviation from geometric simplicity contributes to the radiated emissions potential of the shape.

The measured results are shown in Figure 7.17. The overall result is that the simple square heat spreader shows the lowest potential for radiated emissions, whether grounded or floating. The circular disk is seen to be one of the geometries with the highest potential and also shows a marked difference as to whether the disk is grounded or floated.

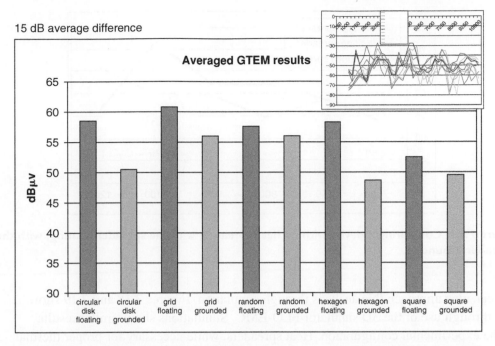

Figure 7.17: Comparison of heat spreader shapes for radiated emissions potential.

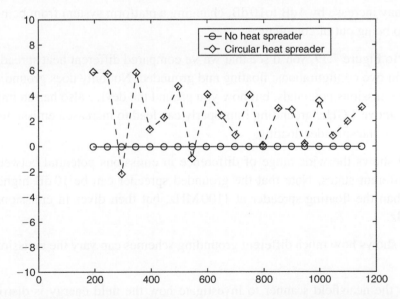

Figure 7.18: Increased emissions due to the presence of a heat spreader.

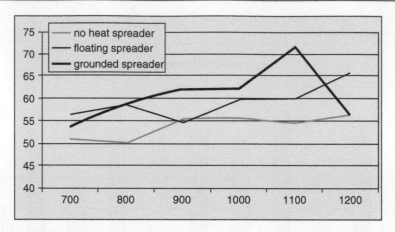

Figure 7.19: Comparing a device with no heat spreader; with the spreader floating; with the spreader grounded.

We can also approach this analysis through simulation. Figures 7.18 and 7.19 show this through use of the 3D simulator FLO-EMC. Simulations show similar results to the experimental configuration. Heat spreaders, while necessary for proper thermal control of high-performance system devices, also need to be properly designed for EMI/RFI considerations. If the radiated emissions potentials are not accounted for, emissions may increase by 4 dB to 10 dB, changing a platform system from being compliant to being out of compliance.

If you refer to Figure 7.17, you'll see that we've compared different heat spreader geometries in two configurations: floating and grounded. Not only does grounding affect the radiated emissions potentials, but how you ground the device also has an impact. At frequencies above 1 GHz, grounding improperly can lead to increased emissions over simply leaving the spreader floating.

Figure 7.19 shows the wide range of difference in emissions potential between the three different states. Note that the grounded spreader can be 10 dB higher in emissions than the floating spreader at 1100 MHz, but then dives in emissions levels at 1200 MHz.

Figure 7.20 shows how much different grounding schemes can vary the emissions for heat spreaders.

We can use the near-field scanner to investigate how the field energy is distributed over the surface of the spreaders. In Figure 7.21, we show two of the structures:

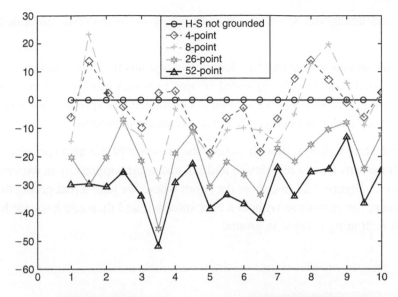

Figure 7.20: Emissions variation due to grounding.

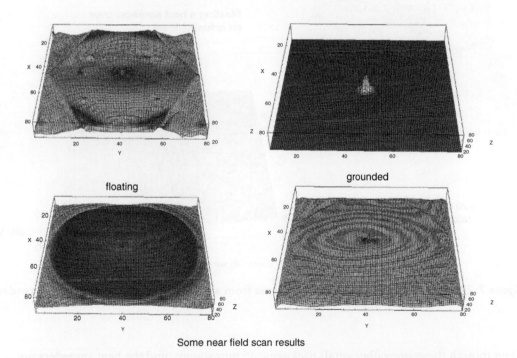

Some near field scan results

Figure 7.21: Near-field scans of the energy distribution across the spreader surfaces.

the circular and hexagonal spreaders. We also show floating versus grounded variations.

For the circular spreader, note that the floating version has the entire surface excited and is a quite good radiator. The grounded version indicates much lower emissions. For the hexagonal spreader, we can see that the grounded hex spreader has a single radiation point located at the center where the excitation is located.

What we would like to know, though, is how the presence of the heat spreader can change the emissions from a real device and not just what happens in an experimental configuration. In Figure 7.22, we show the result where we take a real production chipset, measure the emissions without a heat spreader, and then add a square heat spreader that is floating relative to ground.

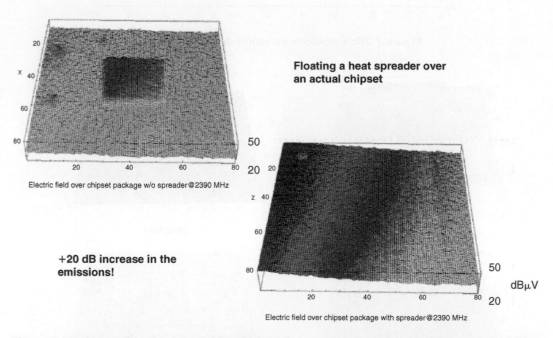

Floating a heat spreader over an actual chipset

Electric field over chipset package w/o spreader@2390 MHz

+20 dB increase in the emissions!

Electric field over chipset package with spreader@2390 MHz

dBµV

Figure 7.22: Measuring the near-field emissions from a chipset with and without the spreader.

Even though these are rather small packages, 35 mm square, and the heat spreaders are somewhat smaller than that, they still become resonant at frequencies around 1.5 GHz (where the first mode occurs). We can calculate the modes for a rectangular shape from the formulas that follow.

$$kx(m) = \frac{m\pi}{\text{length}},$$

$$ky(n) = \frac{n\pi}{\text{width}},$$

$$\text{voltageDist}(x, y) = \cos(kx(m)x)\cos(ky(n)y)$$

$$\text{resonantFreq}(m, n) = \frac{c}{2\pi\sqrt{\varepsilon_r}}\sqrt{\left(\frac{m\pi}{\text{length}}\right)^2 + \left(\frac{n\pi}{\text{width}}\right)^2}$$

7.3 Investigating the Radiated Emissions Potentials of Power Distributions in Packages and Silicon

As Figure 7.23 indicates, there are two major package directions at the present time: wire-bond and flip-chip. In a wire-bond package, the die is facing up and the substrate is in contact with the package substrate upon which it rests. Circuit inputs and outputs are brought into the die on a wire frame. The die is attached to the frame through a wire bond. This is the least expensive package technology and so has been around for a while. Most commodity-level ICs use this package approach. The other direction is called flip-chip. This has the die facing down and the substrate facing up. Connection is made to the PCB and package through solder balls that make direct contact with the package. No wires are used. Because of this contact, density can be significantly increased to a range of 900 to 1200 I/O contacts. This approach also allows for low inductance attachment for the power input.

Figure 7.24 shows the inside of the IC package. Every piece of the package will contribute to the RFI potential of the package. The encapsulation material will have dielectric properties that need to be considered; solder tails that are necessary for fabrication may introduce crosstalk and raise the potential for radiated emissions; and the die attach structure may itself be a good antenna structure. All of these aspects of the package must be considered. In the next sections, we will discuss some of these aspects in greater detail

- There are 2 key paths in package technology:

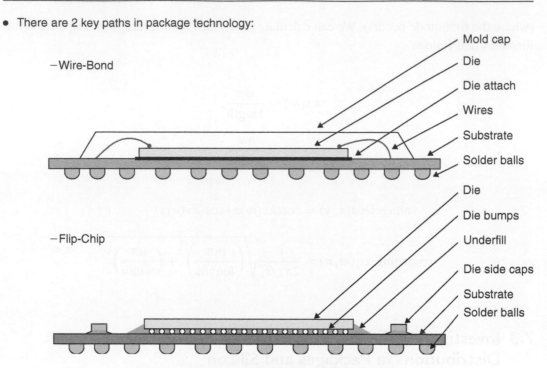

Figure 7.23: Wire-bond and flip-chip packages.

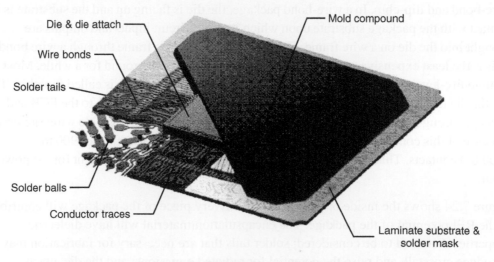

Figure 7.24: Looking inside the package.

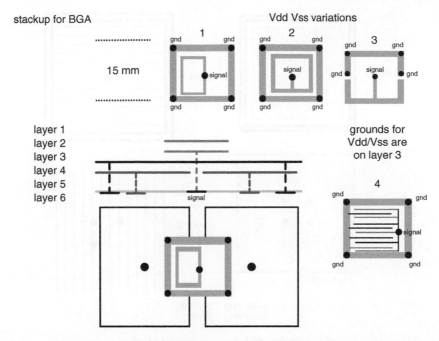

Figure 7.25: Measurement test assembly for emissions testing of power delivery networks.

and outline methods for exploring the RFI impact of different package structures and power distribution both in the package and in the silicon itself.

For our next discussion, we will look into building experimental structures that can mimic silicon power distributions. The test boards have been designed to be used with both the GTEM cell and the Intel NFS system. The test boards come in two parts: The first is 10 cm square, is uniform in topology, and provides a platform for the second part. The second part is a pseudo-BGA meant to simulate the package and the silicon structure. Figure 7.25 shows the general idea for the two parts, and shows the BGA layer stack-up and four of the die power topologies to be investigated. Figure 7.26 shows two additional die geometries that were measured; these two geometries were also used in investigating the effects of swapping the package power layers.

Our intention here is to see if we can abstract the determining geometries of power delivery networks. We'll first look at very simple structures like nested loops, split loops, simple loops, interdigitated nets, and orthogonal nets. We'll describe a GTEM-type test board with a smaller test board that mounts on the larger PCB. The smaller PCB will be used to mimic package and silicon layouts. Figure 7.25 shows a test board that measures

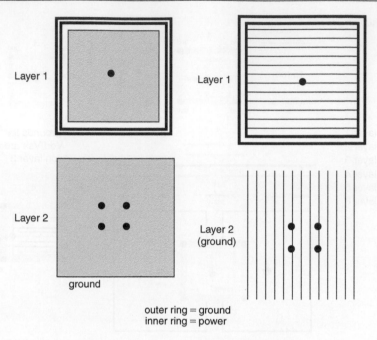

Figure 7.26: Solid versus gridded planes.

the emissions potentials of different package level power delivery geometries. Referring to the results in Figure 7.27:

3PCB_1 corresponds to structure number 1 given in Figure 7.25, two asymmetric loops: This structure was chosen to investigate the effect of asymmetry in co-planar loops.

3PCB_2 corresponds to structure number 2 given in Figure 7.25, two symmetric loops: This structure was chosen to investigate the characteristics of typical die IO power rings.

3PCB_3 corresponds to structure number 3, a single split loop: This structure was chosen to investigate the characteristics of split IO power delivery.

3PCB_4 corresponds to structure number 4, symmetric loops and a gridded array: This simple structure was chosen to investigate the effects of interwoven grids on a single layer.

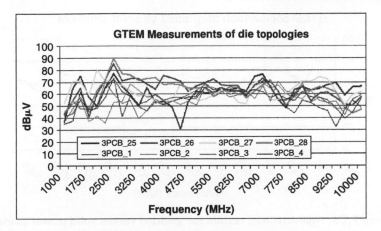

Figure 7.27: Experimental results for the power delivery variations.

3PCB_25 corresponds to the two-layer gridded structure shown in Figure 7.26, with package ground directly below the die layers: This structure was chosen for a preliminary investigation of typical chipset power delivery.

3PCB_26 corresponds to the two-layer solid plane structure shown in Figure 7.26, with package ground directly below the die layers: This structure was chosen to investigate the effect of taking a gridded delivery system to the limit where the power delivery planes become solid and continuous.

3PCB_27 corresponds to the two-layer gridded structure with package Vcc directly below the die layers.

3PCB_28 corresponds to the two-layer solid plane structure with package Vcc directly below the die layers.

As Figure 7.27 indicates, there is quite a spread in emission levels even within such a simple set of power distribution topologies. The highest emission levels were seen with 3PCB_26 and 3PCB_28, when the inner power structure consisted of solid planes. For both gridded and solid distributions, there appeared to be no significant difference attributable to swapping the package power layers. The lowest overall emission levels were seen with the symmetric ring power scheme, 3PCB_2.

Figures 7.28 and 7.29 show GTEM comparisons of gridded versus solid power planes and for swapping power plane positions.

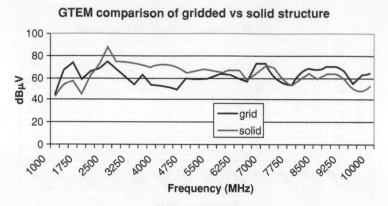

Figure 7.28: Measurement results for gridded versus solid power planes.

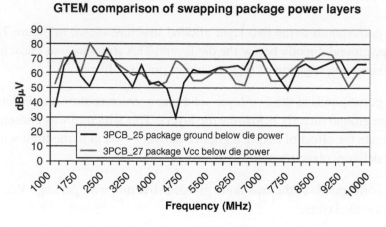

Figure 7.29: Results for swapping power plane positions.

Figure 7.30 is a comparison of the emissions from a two-layer gridded power distribution versus a two-layer solid plane distribution. GTEM measurements indicate that the solid planes produce higher emission levels over a broad frequency range.

Figure 7.29 is a comparison of the emission levels from a two-layer gridded power distribution where the upper package layers are swapped between power and ground being directly beneath the two layers of die power distribution. No significant difference can be seen when the top package layer power planes are swapped.

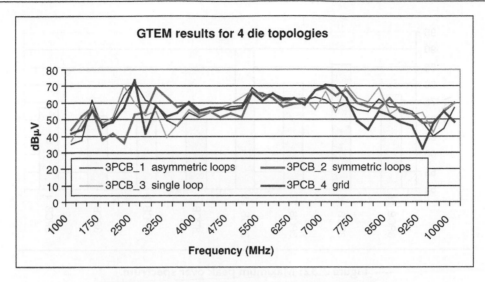

Figure 7.30: Comparison of four simple topologies.

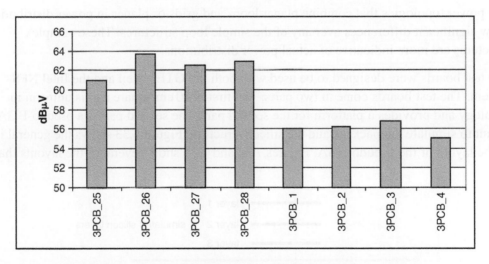

Figure 7.31: Average of maximum emissions over the spectrum.

Comparing both the maximum peaks and the average over the entire spectrum, it is readily apparent that the more complex power topologies have the higher overall emissions. This is shown in Figures 7.31 and 7.32. The simple variations on planar loops show only a several decibel variation in both maximum and average values, though they do show wide variations at any given frequency.

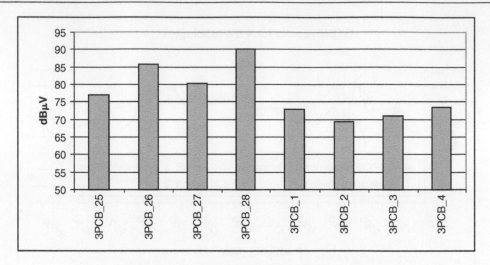

Figure 7.32: Maximum peak over spectrum.

The power topologies that combine planar loops and grids or planes in power distribution show significant differences over any of the simple loop structures. These complex structures are more indicative of actual power distributions.

The test boards were designed to be used with both the GTEM cell and the Intel NFS system. The test boards come in two parts: The first is 10 cm square and is uniform in topology and provides a platform for the second part. The second part is a pseudo-BGA meant to simulate the package and the silicon structure. Figure 7.33 shows the general idea of the layout of the pseudo-BGA. Figures 7.35 and 7.36 show several of the layouts that

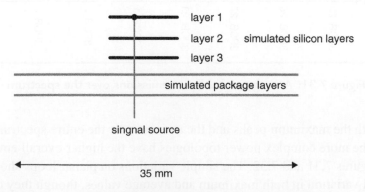

Figure 7.33: Layout of the pseudo-BGA.

were measured. Each die structure is 15 mm square. The power grids were constructed with 4 mil traces.

In an application, layer 1 would be the closest to the substrate.

7.4 Further Investigations with Various Power Topologies

Description of test boards:

- G1S1, G1S2, G1S3: have variations in metal symmetry on the top layer.

- G2S1, G2S2: combine clock and power grids on the same layers.

- G3S1: turns clock net 45 degrees to the power grid.

- G4S2: has interdigitate power and ground on the same layers.

- G6S1, G6S1A, G6S1B, G7S1, G7S1A, G7S1B, G8S1, G8S1A, G8S1B: have variations in clock distribution net and placement in layer stack-ups.

Power delivery in silicon is different from that in packages or motherboards. Due to fabrication constraints, full metal planes cannot be processed. Instead, power is delivered through a complex multilayer mesh of orthogonal nets, as shown in Figure 7.34. We know how power planes radiate and we understand the electromagnetic mechanics of such structures, but how does something like an interwoven mesh radiate? We investigate such

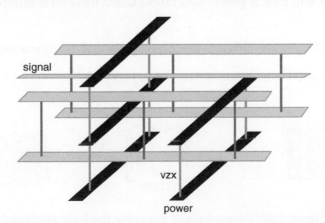

Figure 7.34: Interwoven power distribution mesh typical of silicon power delivery.

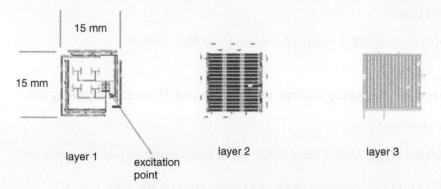

Figure 7.35: An example of the layer stack-ups for one of the clock nets measured.

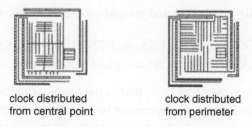

Figure 7.36: The two other clock distributions measured.

structures here. We will look at power structures excited by several different clock tree structures.

Figures 7.35 to 7.37 show the different layer topologies and configurations.

The results of the various topologies are shown in Figure 7.38.

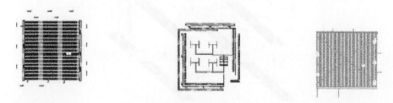

Figure 7.37: An example of swapping the layer stacking.

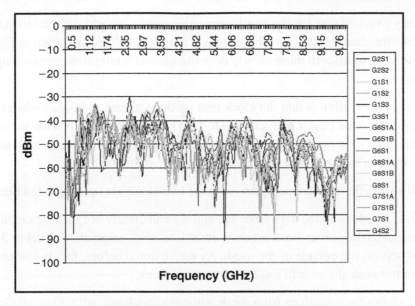

Figure 7.38: Die power delivery emissions.

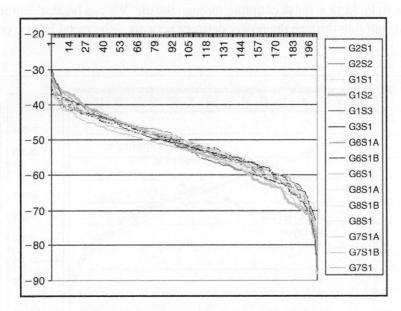

Figure 7.39: Radiated emissions data sorted.

With so many overlaid spectra, the results are rather confusing and difficult to make out. So, we'll take the data sets and sort them from max to min values, as shown in Figure 7.39. This will allow us to discern more clearly how the various configurations stack up in comparison.

An interesting observation is that the clock nets produce higher emissions when they are located deep within the power delivery mesh. This is rather counter to what we may have expected based on experience with motherboards and packages where we seek to locate high-speed circuits between power planes in order to better isolate them.

Figure 7.40 shows the sorted data again, with the two clock net cases pointed out.

In Figure 7.41, we look at the data close up, and we can clearly see that the clock nets located internal to the power mesh produce emissions on average about 2 dB to 3 dB higher than nets on the outside of the mesh. As we've noted before, this is just another knob to turn to tweak the overall total platform emissions.

Figure 7.42 shows the emissions for a mesh without a clock net, with a buried clock net, and with a clock net located on the outside layer of the mesh.

We now turn to looking at noise coupling through the die. We can break the problem into three pieces: coupling through the power delivery network—the metal layers; coupling

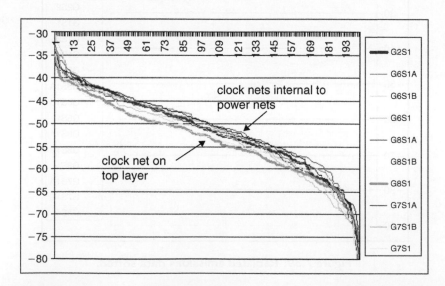

Figure 7.40: Comparison of clock net emissions as a function of layer location in dies.

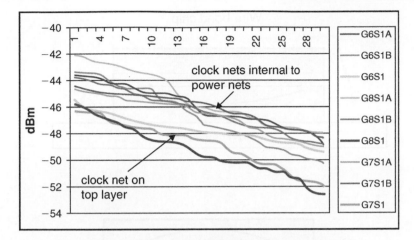

Figure 7.41: Close-up comparison of the emissions.

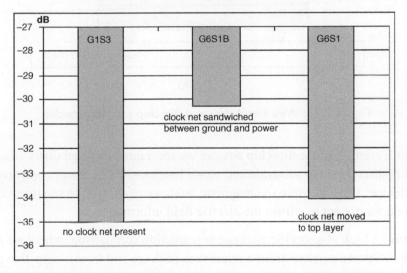

Figure 7.42: Emissions for a mesh without a clock net, with a buried clock net, and with a clock net on the top layer.

through the substrate; and an excitation layer composed of the transistors comprising all the functional areas of the device. This is shown in Figure 7.43.

In Figure 7.44, we show the different near-field energy distributions for flip-chip and wire-bond devices. These are electric field measurements. For the wire-bond device, we see all the detail associated with the metal layers, as they are directly facing the

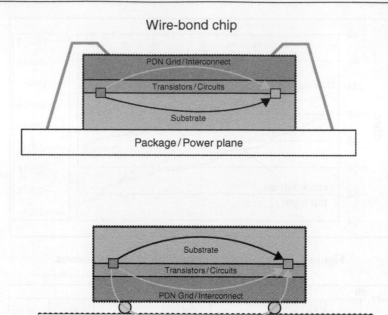

Figure 7.43: Wire-bond chip and flip-chip coupling modes.

measurement system. For the flip-chip device, we see a quite diffused distribution because we're looking directly at the substrate and so we lose most of the detail compared to the wire-bond device. If we measure the magnetic field, we again start to see finer details as the substrate does not act to diffuse the electric field information.

We'll now take a look at a specific package construction comparison, where we can measure the radiated emissions from a microstrip package and from a strip-line package. We will show that the near-field measurement can provide us with quite good information concerning the interior mechanisms that contribute to the radiated emissions.

Also, note that the energy distributions shown here are from real devices, but that they also bear a strong resemblance to the radiation surfaces that we were constructing earlier in the chapter.

In Figure 7.45, we show the measured near-field emissions from the stripline package. Recall that stripline construction has the sources located between two power planes. From our experience, we should expect that this construction will give us the lowest radiated

PDN Grid: differences between flip chip and wire-bond

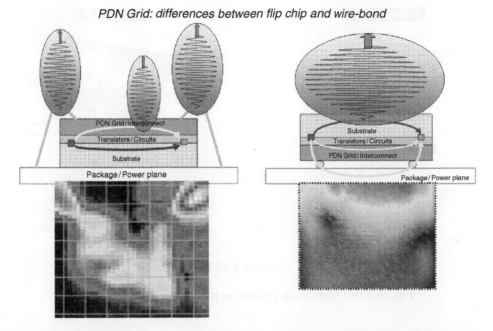

Figure 7.44: Comparing the electric field energy distribution for the two package types.

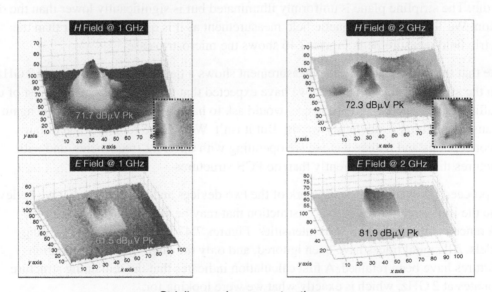

Stripline package construction

Figure 7.45: Stripline processor package measurements.

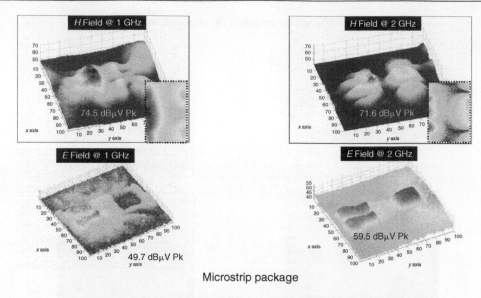

Figure 7.46: Microstrip processor package measurements.

emissions. Note that the electric field level at 2 GHz is 82 dBµV and is centered entirely on the die. The stripline plane is uniformly illuminated but is significantly lower than the die region. We'll ignore the magnetic field measurement as it is quite a bit lower than the electric field measurement. Figure 7.46 shows the microstrip case.

Note that the microstrip package measurement shows a field level 20 dB lower at 2 GHz than the stripline package. Should we have expected that result? I suspect that most of us, recalling our motherboard experience, would ask to have the measurements done again because something is obviously wrong. But it isn't. We need to rethink some of our preconceptions and realize that we are operating with different frequencies, and with structures that respond differently than do PCB structures.

We proceed by examining the artwork of the two devices and extracting what we believe to be the important pieces of the construction that may be the leading causes for the differences in the measured field intensities. Figures 7.47 to 7.49 show the resulting models. All signal traces have been ignored, and only the larger power distribution structures have been retained. A first calculation indicates that the remaining structure resonates at 2 GHz, which is exactly what we were looking for.

After extracting the relevant geometries, we construct the model and use a 3D field solver to produce the predicted field distributions. As Figure 7.49 shows, we have reproduced a

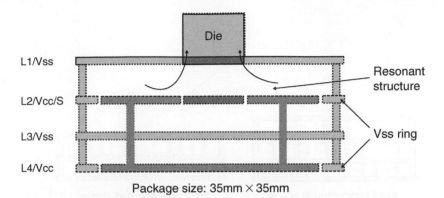

Figure 7.47: Building models to investigate the package differences.

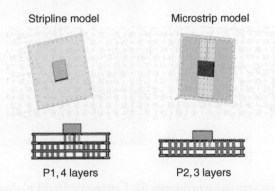

Figure 7.48: Extracting the important pieces of the geometries.

strikingly similar field plot to that which we measured, indicating that our choice of geometry was down the right alley. From this point, we could continue with our investigations and start making modifications to the stripline layout in order to determine whether we could improve on the stripline version versus our microstrip package. In addition, this approach shows good promise in allowing us to design much better mitigation structures in the packages, thereby lowering overall system emissions and interference potentials.

7.4.1 Substrate Noise

We will now turn our attention to mitigation at the silicon level. The trend is to fabricate more and more of the platform systems onto a single substrate creating Systems on Chips.

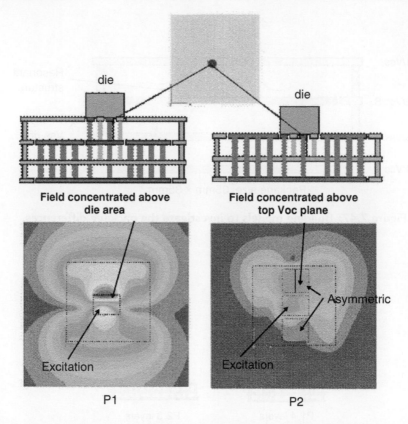

Figure 7.49: Stripline structure produces field intensity 10 dB higher than microstrip.

With high-speed interconnects and low-level sensitive radios sharing the same substrate, the designer needs to know how to specify structures in the substrate that provide adequate isolation between the noise circuits and the potential victim circuits. This is shown in Figure 7.50.

We will now outline a series of measurements that provide us with some guidelines toward how isolation can vary with distance across the substrate—how different guard rings provide different isolation levels. Figure 7.51 shows the baseline structure, which is just a source point and a victim point.

Figure 7.52 shows the results of the measurement made with victim points at varying distances from the source point. The results are pretty straight forward: distance provides higher levels of isolation, which you might have guessed. Also, over a wide frequency range, the variation is reasonably uniform, shown mostly due to measurement.

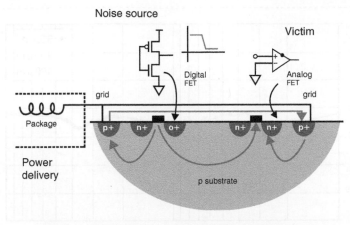

Figure 7.50: Noise coupling in silicon substrates.

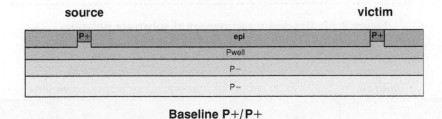

Baseline P+/P+

Figure 7.51: The baseline structure.

What we wish to do is build up silicon structure around the interference source. Figure 7.53 shows the idea where a well is constructed around the source. In this example, the isolation material is composed of "n" material. One can just as well build the structure from "p" material.

Figure 7.54 shows a set of isolation structures. We show where the source alone is isolated, where the victim also has a guard ring, and a P structure guard ring. Other structures are possible, but we'll only indicate what can be done here and we won't go into detail. The comparison chart Figure 7.55 shows what can be accomplished when bias voltages are applied to the guard ring structures.

Guard rings can improve isolation compared to the baseline situation by up to 20 dB to 10 GHz. P material structures show the best isolation values at frequencies below 5 GHz.

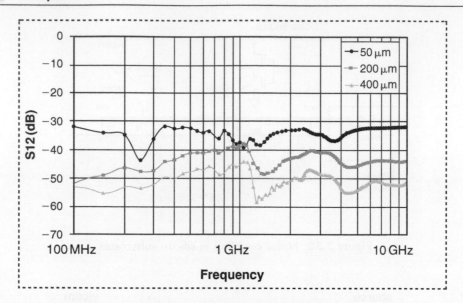

Figure 7.52: Baseline measurement of substrate isolation.

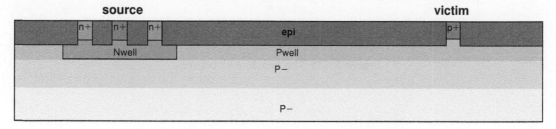

Figure 7.53: Example of N-well guard ring structure.

We will end this chapter by showing how radiated emissions can depend upon where in the process the particular device fell. For the following process, four process corners were identified relating channel size, device size, and so on. We show two steppings of the same processor type and the associated four corners of the process for each stepping. As we can see in Figure 7.56, there can be up to a 10 dB variation across the measured frequency range across the process variations. One must keep this in mind, not only for EMI considerations (where for one corner you might have sufficient margin, but for another if the product fails), but also in terms of interference potentials.

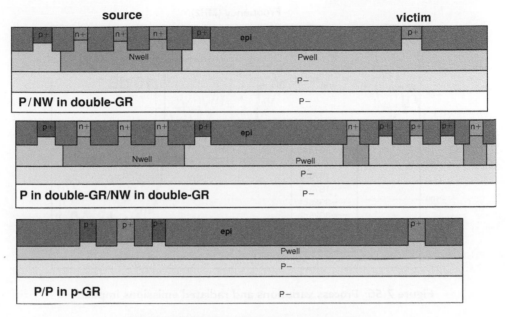

Figure 7.54: Guard ring structures in silicon.

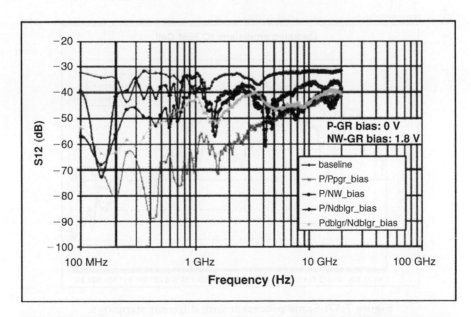

Figure 7.55: Comparison of coupling levels for different isolation structures.

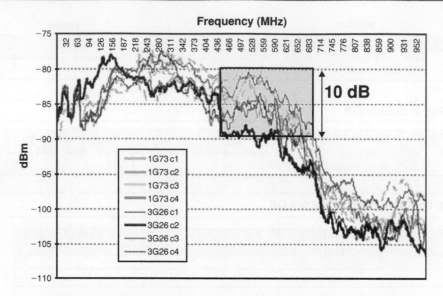

Figure 7.56: Process variations and radiated emissions impact.

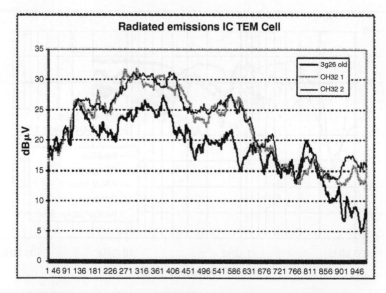

Figure 7.57: Same processor with different steppings.

Figure 7.57 shows the difference in emissions between two steppings of the same processor. Note that in all instances given here, the processors were running the exact same software and were loaded in exactly the same manner.

7.5 Summary

In this chapter, we explored methods for experimenting with PCB construction, connectors, power planes, and silicon and packages. We showed that there are many knobs to turn in order to impact interference. Some knobs provide large changes and some not so many, but they are all usually additive. So as we go through the procedures for mitigation, it isn't necessary that we swing for the bleachers with a home run. Sometimes just hitting those consistent easy to get singles will win the game.

References

[1] F. Melia, *Electrodynamics*, University of Chicago Press, 2001. (Nicely done presentation; the physics is developed along with the mathematics.)

[2] R. P. Feynman, R. B. Leighton, and M. Sands, *Feynman Lectures on Physics*, vol. 2, Addison-Wesley, 2006. (A must-have on every engineer's bookshelf.)

[3] W. A. Blanpied, *Modern Physics: An Introduction to Its Mathematical Language*, Holt, Rinehart and Winston, 1971.

[4] J. Vanderlinde, *Classical Electromagnetic Theory*, 2nd ed., Springer, 2004.

[5] R. E. Hummel, *Electronic Properties of Materials*, 2nd ed., Springer-Verlag, 1993.

[6] C. Paul, *Introduction to Electromagnetic Compatibility*, Wiley, 1992.

[7] C. Paul and S. Nasar, *Introduction to Electromagnetic Fields*, McGraw-Hill, 1987. (Both Clayton Paul books are essential references that should be close at hand.)

[8] S. Ben Dhia, M. Ramdani, and E. Sicard, *Electromagnetic Compatibility of Integrated Circuits*, Springer, 2006.

[9] A. Taflove and S.C. Hagtness, *Computational Electrodynamics*, Artech House, 2005.

[10] B. Sklar, *Digital Communications*, Prentice Hall PTR, 2001.

[11] J. D. Jackson, *Classical Electrodynamics*, Wiley, 1999.

[12] K. A. Milton and J. Schwinger, *Electromagnetic Radiation: Variational Methods, Waveguides and Accelerators*, Springer, 2006.

[13] Jerry P. Marion, *Classical Dynamics of Particles and Systems*, Academic Press, 1965.

[14] K. Slattery, J. Muccioli, and T. North, *Constructing the Lagrangian of VLSI Devices from Near Field Measurements of the Electric and Magnetic Fields*, IEEE 2000 EMC Symposium, Washington, DC.

[15] K. Slattery and W. Cui, *Measuring the Electric and Magnetic Near Fields in VLSI Devices*, IEEE 1999 EMC Symposium, Seattle.

[16] K. Slattery, X. Dong, K. Daniel, *Measurement of a Point Source Radiator Using Magnetic and Electric Probes and Application to Silicon Design of Clock Devices*, IEEE 2007 EMC Symposium, Honolulu.

[17] K. Slattery, J. Muccioli, T. North, *Modeling the Radiated Emissions from Micro-processors and Other VLSI Devices*, IEEE 2000 EMC Symposium, Washington, DC.

CHAPTER 8

Passive Mitigation Techniques

8.1 Introduction

Mitigation techniques can be broken into two categories, which we will call passive and active techniques.

Passive solutions, in a similar way to EMI/EMC shielding or containment, deal with what could be categorized as after-the-fact solutions. These solutions do not generally affect the electronic circuit design. They concentrate on isolating the wireless radio from the interference source. These techniques are somewhat reactive as they deal with reducing coupling with the assumption that the noise is there and the source cannot be addressed any further. Historically, these solutions have been added at the very end of the design cycle. Typically this was the case because of the following:

1. Prediction of EMI/RFI performance was not undertaken, forcing a build-it-and-see approach.

2. These solutions always added cost, so testing was carried out to ensure a minimalist conclusion (enough to meet performance requirements with no room for margin, as margin equates to extra cost).

Following a passive-only approach to mitigation is clearly a risky business. It always adds cost, and the time spent identifying and qualifying fixes in a trial-and-error fashion can easily delay a product for days, weeks, or even months—which in today's environment can be the kiss of death. It is unfortunate that some manufacturers still operate in this mode, planning on four or even five design spins, including last-minute mechanical tweaks.

In order to meet cost and time-to-market (TTM) requirements, a balanced approach must be taken. That takes us to the "active" piece of our mitigation solution space puzzle. These techniques address the actual sources of interference—the moral equivalent of suppression in the EMI/EMC world. This is where the similarity ends, fortunately. Unlike EMI/EMC, where the interference spectrum must be lowered essentially uniformly throughout the entire frequency spectrum, with RFI we are concerned only about the specific frequency bands where the wireless radios exist. This allows an additional degree of flexibility that is not possible with EMI, where noise level reductions in one frequency range can be traded off with noise level increases in bands that do not have any wireless radios. As discussed earlier, wireless sensitivity requirements may be significantly more stringent than EMI requirements, but given this additional flexibility a solution path is indeed possible. Frequency planning in that respect therefore plays a vital role in RFI and will be covered in the pages that follow. Also covered in our active mitigation section will be a segment dealing with frequency content control. Here we will address some not-well-known truths regarding rise/fall times and duty cycles of signals. In this section, we will discuss ways to reduce the frequency content of noise sources without impacting the functional requirements of your system.

It is the intention of the authors to show that by following a balanced and methodical approach to noise mitigation, RFI control can be achieved. By first reducing the noise at the source, setting realistic targets for source to wireless isolation, and then and only then applying passive techniques, the designer can proactively address potential noise issues and negate the need for last-minute "firefighting."

8.2 Passive Mitigation

In this section, we will deal with three different but highly "connected" topics.

1. Shielding

2. Absorbers

3. Layout (including antenna placement)

8.2.1 Shielding

In the extreme case, shielding is sometimes known as a *Faraday cage* or *Faraday shield*. Named after Michael Faraday, who built one in 1836, a Faraday cage or Faraday shield is an enclosure formed by conducting material, or by a mesh of such material. Such an enclosure blocks out external electromagnetic fields or contains internal electromagnetic fields.

Two approaches are generally taken to the creation of such cages in product development.

1. Follow/Copy what was done before and was good enough on the previous design, and keep your fingers crossed that the electronics didn't change too much.

2. Start from scratch. Construct something as physically close to a sealed metal box as possible, perhaps even throwing everything you can at it as if you were burned in a previous life by EMI/RFI. Test it, tweak it, and follow general trial-and-error problem solving until it passes EMI/TIS (total isotropic sensitivity) testing.

Given the strides made in simulation tools and the computer resources at the fingertips of today's engineers, this would seem to be a questionable approach, especially since it almost always ends up costing time and money as previously indicated.

So what is a better way to ensure that your design incorporates what is necessary and sufficient to meet your product performance requirements?

In Chapter 1, we asserted that the RFI/EMI engineer needs to be involved in every aspect of the design, from the selection and layout of electronic components to the physical attributes of the PCB to the mechanical design of the enclosure. To design an effective shield, the responsible engineer must first have an inherent knowledge of the complete design.

Armed with this knowledge, one should take the following approach:

1. Set a shielding/isolation target.

    *   For the targeted radios, determine a noise threshold level.

    *   For the possible noise sources within the system, determine through measurements on previous systems or simulation data the predicted magnitude of platform-generated noise in the frequency bands of interest.

    *   Using the above information, determine the isolation required between the noise source and the wireless radio.

2. Establish factors that will affect achievable shielding effectiveness (SE).

    Let's start by defining SE.

    For *E* fields

    $$\mathrm{SE} = 20 \log E_{\mathrm{in}} / E_{\mathrm{out}} \; (\mathrm{dB}).$$

    For *H* fields

    $$\mathrm{SE} = 20 \log H_{\mathrm{in}} / H_{\mathrm{out}} \; (\mathrm{dB}).$$

If shields were perfect, E_{out} and H_{out} would be zero. In reality, a shield is an attenuator that performs based on two fundamental principles: reflection and penetration loss.

Penetration (absorption) loss increases with

- Thickness

- Frequency

- Conductivity

- Permeability

Reflection increases with

- Surface conductivity

- Wave impedance

We will first look at penetration loss. To understand penetration loss (also known as absorption loss), we'll first discuss the concept of skin depth.

When an electromagnetic wave passes through a conductive material, its amplitude decreases exponentially due to ohmic losses. This can be written as

$$E_{out} = E_{in}e^{-t/\delta} \quad \text{and} \quad H_{out} = H_{in}e^{-t/\delta},$$

where E_{in} and H_{in} are the wave intensities at the surface of the material and E_{out} and H_{out} are the relative wave intensities at a distance t inside the material. The distance for the wave to be attenuated to $1/e$ or 37% of its original value is defined as the skin depth. This is shown in Figure 8.1 as an exponentially decreasing current density with each point illustrating 1 skin depth.

For a given material, the skin depth can be calculated from

$$\text{Skin depth } \delta(m) = \frac{1}{\sqrt{\pi f \mu \sigma}},$$

where

f = frequency in Hz
μ = permeability
σ = conductivity

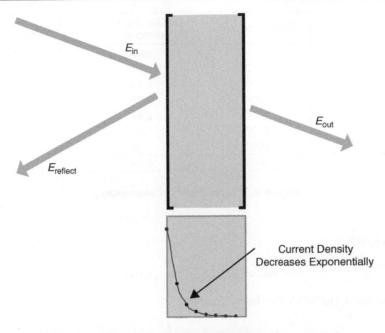

Figure 8.1: Simple depiction of shielding characteristics.

Substituting numerical values, this can be rewritten as

$$\text{Skin depth } \delta(m) = \frac{1}{15.158\sqrt{f\mu_r\sigma_r}}$$

where μ_r and σ_r are the relative permeability and conductivity with respect to copper.

Note that the skin depth decreases with the square root of the frequency. Physically, this means that as the frequency of the current flowing in a conductor is increased, the majority of the current tends to crowd toward the outer surface of the conductor or, in our case, the shield.

To illustrate this, Figure 8.2 shows the calculated skin depth for aluminum ($\sigma_r = 0.61$) at various frequencies.

It is therefore clear that skin depth is an important factor in determining absorption loss. In fact, in order to calculate the absorption loss, we need to know how many skin depths the shield material represents at the frequencies of concern, since the field strength will reduce by ~8.7 dB (or 63%) for every skin depth.

Frequency (MHz)	Skin Depth (microns)
1	84.5
10	26.7
100	8.45
500	3.78
1000	2.67
3000	1.54
5000	1.19
10000	0.84

Figure 8.2: Skin depth of aluminum.

As stated previously,

$$SE = 20 \log E_{in}/E_{out} \text{ (dB)}.$$

Given that, the wave intensity can be represented by

$$E_{out} = E_{in}e^{-t/\delta}$$

$$\frac{E_{in}}{E_{out}} = \frac{1}{e^{-t/\delta}} = e^{t/\delta}.$$

So, for our absorption loss SE_A, substituting in the equation above, we get

$$SE_A = 20 \log \left(e^{t/\delta}\right)$$

$$SE_A = 20 \left(\frac{t}{\delta}\right) \log e$$

$$SE_A = 8.69 \left(\frac{t}{\delta}\right)$$

Replacing $\delta = \dfrac{1}{15.158\sqrt{f\mu_r\sigma_r}}$ in the above equation, we can therefore come to a relatively simple expression for absorption loss (t has been translated from meters to mm and F from Hz to MHz to ensure that variables are represented in appropriate units).

$$SE_A(\text{dB}) = 131 \times t \times \sqrt{F\mu_r\sigma_r},$$

where

t = thickness of shield material in mm
F = frequency in MHz

μ_r = permeability relative to copper
σ_r = conductivity relative to copper

For example, a 0.01 mm (10 micron) thick aluminum shield will offer an absorption loss at 500 MHz of:

$$A_{dB} = 131 \times 0.01\sqrt{500 \times 1 \times 0.61} = 22.9\,dB.$$

Absorption loss versus frequency for two thicknesses of copper and aluminum is shown in Figure 8.3. As we can see, absorption loss increases with frequency and with material thickness.

For reference purposes, we have included Figure 8.4, which details the characteristics of some common shielding materials.

For nonmagnetic materials ($\mu_r = 1$), the absorption losses increase with conductivity (σ_r). Given that silver is the only metal that offers better conductivity than copper, and silver is somewhat cost-prohibitive, it can be generalized that any non-magnetic material will show less absorption loss than copper. For example, nickel, with a similar thickness in the example above (10 micron), which has a conductivity $\sigma_r = 0.2$, would have the following absorption at 500 MHz.

$$A_{dB} = 131 \times 0.01\sqrt{500 \times 1 \times 0.2} = 13.1\,dB.$$

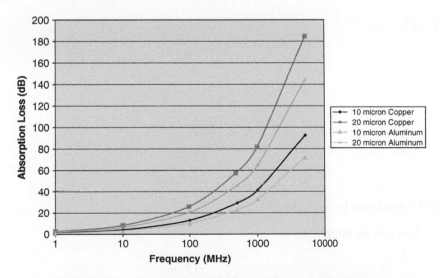

Figure 8.3: Absorption losses of copper and aluminum.

Metal	Conductivity Relative to	Relative Permeability	Absorption Loss (dB per micron = 0.001mm)		
			1MHz	100MHz	1GHz
Silver	1.05	1	0.13	1.34	4.24
Copper- Annealed	1	1	0.13	1.31	4.14
Copper- Hard Drawn	0.97	1	0.13	1.29	4.07
Gold	0.7	1	0.11	1.1	3.46
Aluminum	0.61	1	0.1	1.02	3.23
Magnesium	0.38	1	0.08	0.8	2.55
Zinc	0.29	1	0.07	0.71	2.23
Brass	0.26	1	0.07	0.67	2.11
Cadmium	0.23	1	0.06	0.63	1.99
Nickel	0.2	1	0.06	0.59	1.85
Bronze	0.18	1	0.06	0.56	1.76
Iron	0.17	1000	1.71	17.08	54.01
Tin	0.15	1	0.05	0.51	1.6
Steel (SAE 1045)	0.1	1000	1.31	13.1	41.43
Beryllium	0.1	1	0.04	0.41	1.31
Stainless Steel	0.02	~1	0.02	0.19	0.59

Figure 8.4: Characteristics of metals used for shielding.

We will now consider the reflection component. Here things get a bit trickier because the reflection component will be different if the shield is in the near field or far field. For the far field, we can neglect internal reflection. Therefore, the loss can be calculated by:

$$R_{dB} = 20 \log \frac{(K+1)^2}{4K}$$

where

$$K = \frac{120\pi}{Z_s}$$

and

Z_s = shield impedance in ohms/square.

For $K > 3$, this can be simplified to

$$R_{dB} = 20 \log \frac{120\pi}{4Z_s}.$$

To calculate the reflection loss, we must therefore know the Shield impedance Z_s. The characteristic impedance of a medium is defined as

$$Z = \sqrt{\frac{j\omega\mu}{\sigma + j\omega\varepsilon}}$$

where σ = dielectric constant.

For conductors where $\sigma \gg j\omega\varepsilon$, the characteristic impedance is known as the *shield impedance* and can be simplified to

$$Z_s = \sqrt{\frac{j\omega\mu}{\sigma}} = \sqrt{\frac{\omega\mu}{2\sigma}}(1+j).$$

Considering only the magnitude, we therefore arrive at

$$|Z_s| = \sqrt{\frac{\omega\mu}{\sigma}} = \sqrt{\frac{2\pi f \mu}{\sigma}}.$$

Substituting numerical values, we can generalize this equation for any conductor to

$$|Z_s| = 3.68 \times 10^{-7}\sqrt{\frac{\mu_r}{\sigma_r}}\sqrt{f}.$$

Plugging in the values for copper and aluminum, we therefore get

$$\text{Copper } |Z_s| = 3.68 \times 10^{-7}\sqrt{f}$$

$$\text{Aluminum } |Z_s| = 4.71 \times 10^{-7}\sqrt{f}.$$

Plugging the generic equation for any conductor into our reflective loss equation, we can now create a general expression.

$$R_{dB} = 20 \log \frac{120\pi}{4Z_s}$$

$$= 20 \log \frac{30\pi\sqrt{\sigma_r}}{3.68 \times 10^{-7}\sqrt{f\mu_r}}$$

$$= 20 \log \frac{2.56 \times 10^8 \sqrt{\sigma_r}}{\sqrt{f\mu_r}}$$

Using the expressions, we can now plot both the absorption and reflective losses. Such plots for a 10 micron thin copper shield are shown in Figure 8.5. Also shown is the cumulative SE.

In Figure 8.5, it is clear that the reflective loss decreases linearly with increasing frequency (log scale), the absorption loss increasing somewhat exponentially over the same range.

It is also apparent that even very thin copper can be a good shield in the far field. Unfortunately, for the subject matter in this book, given the frequencies of interest and the likely form factors of the final product, it is clear that we cannot make this assumption. As an example, we will look at 2.5 GHz, where some of today's WLAN radios operate. To calculate the wavelength, we use the following equation.

$$\lambda(m) = \frac{300}{F(\text{MHz})}$$

So, at 2.5 GHz, the expression is as follows.

$$\lambda(m) = \frac{300}{2500} = 0.12\,\text{m} = 12\,\text{cm}.$$

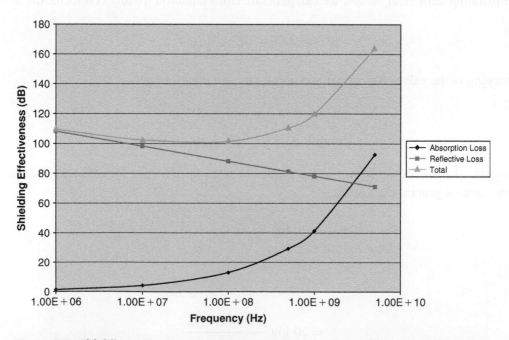

Figure 8.5: Shielding effectiveness components for 10 micron thin copper (far field).

There is much debate about where the transition from near field to far field occurs. For the purpose of this illustration, we will use $\lambda/6$. For our frequency of 2.5 GHz, that would equate to 2 cm, just a little shy of 1 inch. If you look at most mobile devices today, it is clear that there is nothing like this distance between the electronics and the shield. Therefore, for at least these radios we need to assume that our shield will firmly be in the near field. Now that we've established that the shield is in the near field, we have the added complexity that the reflection loss is dependent on whether the noise is electric or magnetic in nature.

We will now look at the expressions for the near-field cases. Firstly we will deal with the E field.

For E fields,

$$R_{dB(E)} = 20 \log \left[\frac{120\pi}{4Z_s} \times \frac{\lambda}{2\pi D} \right]$$

Far-field reflection term

Near–to-far field correction

$$= 20 \log \frac{4500}{DFZ_s} \quad \text{[Replacing } \lambda = \frac{300}{F} \text{ in the equation]}$$

Substituting a general expression for Z_s and making the units consistent, we get

$$R_{dB(E)} = 20 \log \frac{1.22 \times 10^{16} \sqrt{\sigma_r}}{Df \sqrt{\mu_r f}},$$

where

f = frequency (Hz)
D = distance from the noise source (meters)
μ_r = permeability relative to copper
σ_r = conductivity relative to copper

For H fields,

$$R_{dB(H)} = 20 \log \left[\frac{120\pi}{4Z_s} \times \frac{2\pi D}{\lambda} \right]$$

Far-field reflection term

Near–to–far field correction

$$\sim= 20 \log \frac{2DF}{Z_s} \quad \text{[Replacing } \lambda = \frac{300}{F} \text{ in the equation],}$$

where

Z_s = shield barrier impedance (Ohms/square)
F = frequency (MHz)
D = distance from noise source (meters).

Substituting a general expression for Z_s and making the units consistent, we get

$$R_{dB(H)} = 20 \log \frac{5.43 Df \sqrt{\sigma_r}}{\sqrt{\mu_r f}},$$

where

f = frequency (Hz)
D = distance from the noise source (meters)
μ_r = permeability relative to copper
σ_r = conductivity relative to copper.

Using these equations, we can now plot the reflective loss for both magnetic and electric fields in the near field. This is shown in Figure 8.6, where the reflective loss for a copper shield 1 cm away from a noise source is plotted. The far-field case is also plotted for completeness. Notice that the electric and magnetic losses converge until they are equal at a distance which equates to $\frac{\lambda}{2\pi}$. Beyond this, the near-field equations should not be used, as the reflective loss for both magnetic and electric fields equates to that of a plane wave, so the far-field equation for reflective loss is applicable. The far-field reflective loss is plotted for reference purposes only and should be ignored in this case for frequencies below the near-field to far-field transition, which occurs at $\frac{\lambda}{2\pi}$.

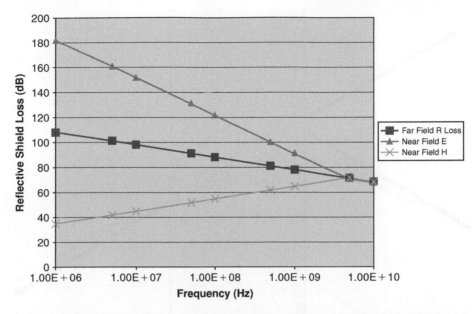

Figure 8.6: Reflective loss plots for a copper shield 1 cm from a noise source.

The obvious problem is that even if you know the equations, how do you figure out if your noise source is predominantly magnetic (H) or electric (E). In the preceding chapters, we introduced some measurement methodologies where this information can perhaps be garnered. The near-field scanning (NFS) system, which uses both electric and magnetic probes to measure noise, can clearly provide this information. Looking at the type of noise source can also help to figure out the major mode. Circuits with large switching currents (such as power supplies or voltage regulator circuits) tend to generate magnetic fields, where high-impedance voltage mode circuits like those used for high-speed I/O tend to create electric fields.

To ensure completeness, one other reflective loss factor should be mentioned: That is the concept of a correction factor for the impact of internal reflections at the boundary walls. This is depicted in Figure 8.7.

If the shield is thin, the reflected wave from the boundary at the other side of the shield (annotated as the "reflected absorption wave" in Figure 8.7) is re-reflected off the incident boundary wall, where it is reflected again, and so on.

As Figure 8.7 shows, the majority of the electric field energy is reflected off at the incident boundary, so with little energy finding its way into the shield we can neglect this effect for

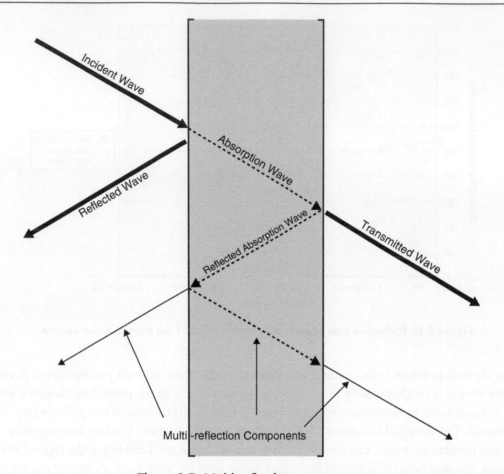

Figure 8.7: Multi-reflection components.

electric fields. Unfortunately, for magnetic fields, the opposite is true; most of the energy enters the shield. To calculate this correction factor for magnetic fields, the following expression can be used.

$$R_{MR(H)} = 20 \log\left(1 - e^{-2t/\delta}\right),$$

where

t = shield thickness

σ = skin depth.

This is plotted in Figure 8.8.

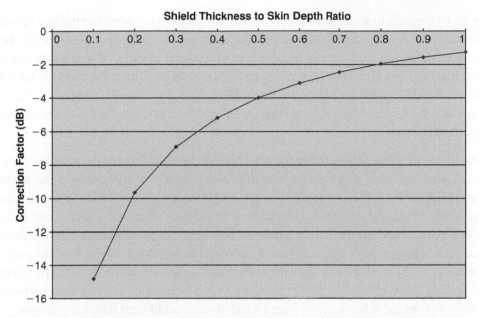

Figure 8.8: Correction factor for magnetic fields with thin shields (< skin depth).

We should first notice that the correction factor is actually negative, which means that the overall shielding is reduced by this multi-reflection effect. Secondly, we can see that the correction factor decreases as the shield thickness approaches 1 skin depth, and only equates to ~1 dB at 1 skin depth. Given the frequencies of interest for wireless interference (> 500 MHz) and that it is unlikely that any shielding solution would be less than 1 skin depth thick, we can therefore ignore the multi-reflection component in our calculations. In fact, this effect can generally be neglected in the case of shields at least 1 skin depth thick, given that by the time the wave reaches the outer boundary (having passed through 3 skin depths), it is of negligible amplitude.

So it would seem that shielding should be a reasonably easy approach to providing isolation between your noisy system components and your wireless components. There are two further aspects of shielding that we should consider before we can draw such conclusions.

The best shielding solution would be a perfect seamless enclosure made of copper, silver, aluminum, nickel, zinc, or even gold. Unfortunately, this lofty goal is impractical, cost-prohibitive, and, most importantly, integrated antennas would not function (given that they would be inside the shield). Not to mention any required system cooling with airflow needs may be tough to implement given the lack of holes. Also note that the use of metal in the enclosure design may be impossible due to shape, weight, or cost constraints. In these cases, designers will therefore make extensive use of plastics or specifically injection molded plastics. Since plastics generally provide no shielding, metallic coatings have been developed to add shielding characteristics to these plastic enclosures. Given that these coatings will be thinner than conventional shielding materials, we need to comprehend this in our calculations.

In view of the fact that material thickness is more important at lower frequencies (skin depth is larger, so fields penetrate more at low frequencies), we will consider the worst case as a mobile TV receiver operating in the 500 MHz range. We'll then calculate some requirements for shield thickness. For example, if we take aluminum paint as our material of choice, we can see from Figure 8.2 that the skin depth of aluminum is 3.78 microns. Earlier in the chapter, we calculated that the absorption loss for a 10 micron thin piece of aluminum is ~22 dB. If we calculate the reflection loss for aluminum at 500 MHz, we get 79 dB making a total SE of ~100 dB. So, even very thin metallic coatings can perform very well. For added reference, we have included some additional data for common coatings.

(All of these data represent typical values. For accurate values, contact the manufacturer of a specific coating.)

Graphite: The least conductive of the common shielding coatings at 30–50 ohms/square/mil. Provides 30–50 dB of protection along a range of 50–450 MHz at 2 mils coating thickness.

Nickel: Excellent durability due to hardness, along with low cost, make nickel a good choice if conductivity between copper and graphite is called for. Conductivity is around 1 ohm/square/mil, which provides as much as 60–65 dB of protection at 2 mils over 5–1,800 MHz.

Copper: Probably the most significant commercially-used shielding coating material; many copper-based coatings use silver-plated copper fillers to improve resistance to oxidation of the copper. The conductivity of the copper shielding coatings is significantly greater than the previous two types: 0.075 to 0.10 ohms/square/mil. Copper coatings offer 75 dB to over 1 GHz of protection.

Silver: The most conductive of commonly used coatings, silver's high cost prohibits its use in all but the most demanding circumstances. Outstanding conductivity of 0.010 ohms/square/mil provides effective shielding up to 10 GHz at over 75 dB.

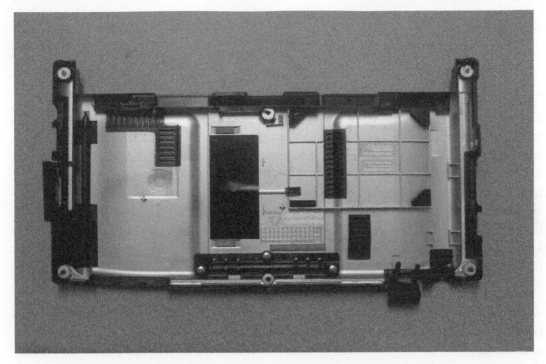

Figure 8.9: Production example of a metallic coating for shielding.

Figure 8.9 shows an example of a production system using a metallic coating as a shielding solution.

In this chapter, we have shown how to calculate the SE of essentially a perfect shield, based on material choice, material thickness, locality to the noise source (near-field or far-field), and the characteristics of the noise itself. Unfortunately, as mentioned earlier in the text, there are many reasons why an essentially contiguous shield cannot be created. These reasons can range from thermal cooling requirements to accessibility for customer or field upgrades to mechanical joints and seams, or, most likely in the wireless world, to openings for antennas. All of these reasons considerably reduce the effectiveness of the shield. When it comes down to it, the intrinsic SE of the material is of little concern compared to the leakage through these holes, antenna openings, or mechanical joints and seams. These features represent what are known as *shield discontinuities*. The amount of leakage though these discontinuities is defined by three main factors.

1. Maximum dimension of the aperture

2. Frequency of the noise source

3. Wave impedance of the source

To understand how openings affect shielding, we will consider the physics of an electromagnetic wave impinging on a conductive surface. When the wave hits the surface, it will generate currents that flow along the surface of the material. These currents create a reflected wave of opposite polarity to the incident wave, leading to cancellation of the incident field (this is a requirement to satisfy the well-known boundary condition that the total electric field tangential to a perfect conductor must be zero). For this to happen, the currents must flow unhindered, or cancellation will not occur. If we now consider a hole or opening in the shield, it is clear that the surface currents will no longer be unhindered, and in a similar fashion to a split in a ground plane forcing a return current along a less than ideal path (not the shortest), our surface current will be forced to take a longer path. This is illustrated in Figure 8.10.

Also illustrated in Figure 8.10 is that fact that the larger the slot, the greater the perturbation of the currents and a greater reduction in shielding. From this figure, we can also see how it is the largest dimension that matters as we could in fact double the area of the slot by increasing the vertical dimension (and the current disruption would not increase significantly). If we also consider a large amount of small holes, it is clear that the resultant perturbation would be less than a single large hole. Slots such as those shown in the figure can easily form what is known as a *slot antenna*. This is good if you wish to transmit though such a gap, but bad if it was not your intent.

For slots equal to less than $\lambda/2$ (one-half wavelength), the SE can be expressed as

$$S = 20 \log\left(\frac{\lambda}{2l}\right),$$

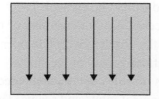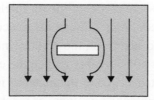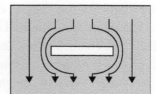

Figure 8.10: Effects of slots on surface currents: unhindered surface currents, small slot impact, and large slot impact (*left* to *right*).

where

λ = wavelength

l = maximum dimension of the aperture.

Substituting $\lambda = 3 \times 10^8 / f$ (f = frequency in Hz), the above expression is rewritten as

$$S = 20 \log\left(\frac{3 \times 10^8}{2lf}\right).$$

When we look at the first equation above, it is clear that when the aperture dimension is $\lambda/2$, the SE is 0 dB. It is also easy to see from the expression that the SE increases by 6 dB every time the slot size is reduced by a factor of one half and that the SE increases by 20 dB/decade as the aperture size is decreased. This is shown in Figure 8.11.

The plot in Figure 8.11 represents the SE for various sizes of aperture as a function of frequency. Rarely is there just one aperture, and more than one will decrease the SE yet again. The reduction in SE due to multiple apertures or slots will depend on a number of factors including the distance between the slots and the number of apertures. As a general

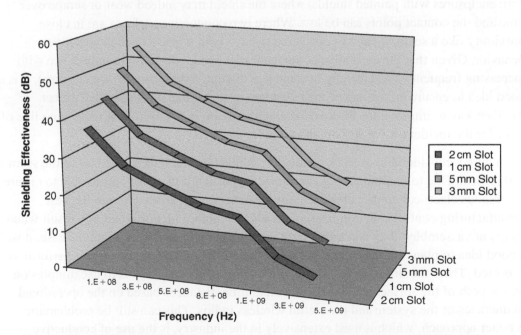

Figure 8.11: Shielding effectiveness vs. frequency for various aperture geometries.

rule, when apertures of equal size are placed closer together than $\frac{\lambda}{2}$, the reduction in SE can be described by

$$S = -20 \log \sqrt{n},$$

where n = number of apertures.

We should note that this only applies to holes on the same surface. Holes located on other enclosure surfaces will not collude to reduce the SE any further. This being the case, it is a good idea to distribute any required apertures throughout all surfaces of the equipment, thus minimizing the impact from multiple apertures.

As a general rule of thumb, aperture sizes should be kept to somewhere between $1/15\lambda$ and $1/20\lambda$ to ensure approximately 20 dB of SE. This would need to be reduced accordingly if multiple openings are required.

As it turns out, holes or apertures for thermal/cooling purposes tend to be very controlled. That cannot be said of seams or mechanical edges; they tend to act as a series of slots with scattered contact points along their length. Unfortunately, these contact points can be inconsistent and can vary when the equipment is assembled. This is particularly an issue with enclosures with painted shields, where the shield may indeed wear or scrape over time and the contact points can be lost. Where two conductive surfaces are in close proximity like a seam, it can be considered to have both a resistive and capacitive behavior. Given that the capacitive component will yield a decrease in impedance with increasing frequency (and thereby become less disruptive to those surface currents), it is a good idea to ensure an overlap of the two edges to maximize this capacitive aspect. Another way to improve the SE associated with a seam is to ensure that there is no line of sight for the incident wave. This is shown in Figure 8.12.

Even with a considerable overlap, conductive contact between the two edges of the seam is still critical. Due to manufacturing variations, it is difficult or nearly impossible to ensure continuous electrical contact along lengthy straight contact areas. Even in the best manufacturing controls environment, materials are seldom identical and as a result when pieces are assembled they don't always fit together as intended. This being the case, it isn't a good idea to plan on incidental electrical contact in your design, as it won't perform as expected. There are different ways to deal with this issue. One way is to use dimples on one or both of the surfaces at appropriate separation distances based on the operational frequencies of the system and the victim wireless radios. This can still be problematic. A better approach, which is used extensively in the industry, is the use of conductive gaskets or elastomers. These gaskets are intended to take up any manufacturing tolerance

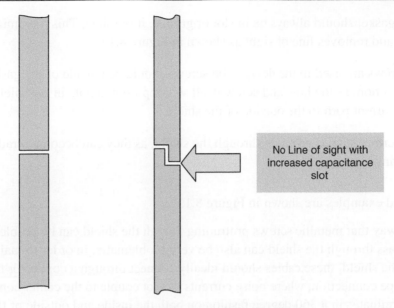

No Line of sight with increased capacitance slot

Figure 8.12: Improving SE of seams.

and ensure a good conductive contact between the two pieces of the shield. This does add cost over the simple mechanical dimples but can yield significantly better results. Even with these gaskets, care should be taken as their performance depends on two things: their conductivity and their composition relative to the intended metal surfaces.

Conductivity: Proper installation/implementation of these gaskets requires compression. As a general rule, the greater the compression the lower the resultant impedance associated with the metal-to-gasket-to-metal surface and the gasket-to-gasket connections. If these gaskets are glued to one of the surfaces, it's a good idea to ensure that the glue is conductive. It may seem common sense, but given conductive glue is generally more expensive than common glue, beware of the cost-reduction engineer who comes along and negates all your engineering work to a save a fraction of a cent.

Galvanic compatibility: Electrolytic corrosion (electrolysis) also known as *galvanic action* occurs when dissimilar metals are in contact in the presence of an electrolyte, such as water (moisture). The dissimilar metals set up a galvanic action, resulting in the deterioration of one of them. This being the case, the designer needs to be diligent in selection of gasket material to ensure that the gasket and the intended metal mating surface are *galvanically compatible*.

There is, of course, a correct and incorrect way of implementing a gasket solution.

1. The gasket should always be in slot or groove, if possible. This maximizes contact area and removes line of sight as shown in Figure 8.12.

2. If screws are used in the design, the screws should be outside of the gasket connection, as the hole and screw itself will represent a hole in the shield as well as a current path to the outside of the shield.

3. No screws should protrude through the shield, as they can become reradiating antennae.

Good and bad examples are shown in Figure 8.13.

In the same way that metallic screws protruding though the shield can be problematic, cables that pass through the shield can also be very problematic. In order to maintain the integrity of the shield, these cables should ideally connect through a connector that allows for a coax-type connection, where noise currents cannot couple to the center conductor and the shield terminates in a 360-degree fashion on both the inside and outside of the enclosure. If this cannot be achieved, it is vitally important to provide contingency for electrical filtering of the signals so that any high-frequency noise coupled onto the signal can be filtered prior to leaving the enclosure. This filtering should happen as close to the connector as possible, within the connector itself if at all possible.

Although not typically associated with the subject of shielding, one aspect that needs to be addressed is antenna and shield interactions. It is very common for RF designers to connect the antenna return directly to nearby shields in the system. This is done to provide a low impedance return path for the antenna, improving the performance/coverage of the antenna. It turns out that for noise, this may not be the best approach. If noise currents are flowing on the shield, these currents will pass straight into the front end of the radio and cause a multitude of problems. Wireless designers will argue that they need this connection to ensure sufficient radio performance. Unfortunately, the reality is that in the majority of cases where "noise" is not considered, the radio benefit gained may be more than lost, the noise currents resulting in a reduced signal-to-noise ratio (SNR). We will cover this topic in more detail later in the chapter when we talk about antenna placement as a potential weapon in our arsenal against platform interference.

We will now look at a real-world example of shielding. In Figure 8.14, we see multiple examples of shielding as a control mechanism. There are multiple pieces over particular areas—conductive gaskets, and shielding tapes. In this example, it would seem that

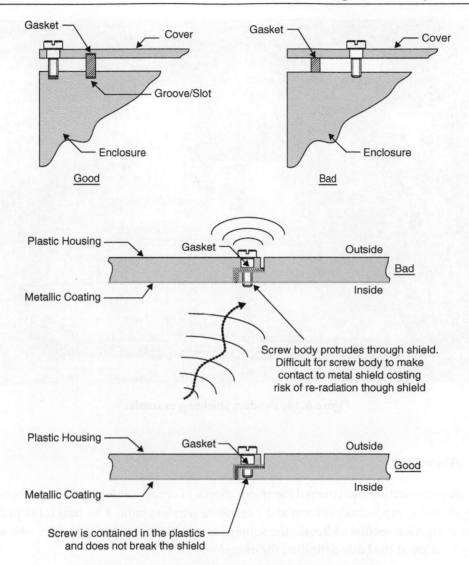

Figure 8.13: Gasket design examples (good and bad).

at least some of their solutions are what are called *band-aids*. Therein lies the biggest disadvantage of a predominately shielding approach; more often than not, last-minute tweaks or band-aids will be required. Also, in this picture we will see the selective use of absorber materials (over cable for a storage device), which is the next topic we will cover.

Figure 8.14: Product shielding example.

8.2.2 Absorbers

In the previous section, we covered the major pieces to create a substantial shield between a noisy digital components/platform and a sensitive wireless radio. Our next topic in the passive mitigation section addresses the somewhat symbiotic topic of absorbers. We will start by looking at the basic definition for *absorb*.

Ab·sorb (ab-sôrb', -zôrb'): 1. To take in; soak up 2. To take in and assimulate: Plants absorb energy from the sun 3. To take in or receive (e.g. sound) with little or none of it being transmitted or reflected.

The third part of the definition is the key here. In the shielding section, we talked about the absorption and reflective components in calculating the total SE. Normal metallic materials have an absorption loss but tend to reflect considerable amounts of energy, depending on their conductivity. Absorbers have the key aspect that they do not reflect any significant levels of energy due in part to their low conductivity. Unlike isolation materials (which are essentially invisible to EM waves and therefore the energy passes though unobstructed and unattentuated), absorbers allow the wave to pass though but attenuate the wave as it does so. They are essentially shields without the reflective loss component. This is achieved by doping a host material, generally a common dielectric material such as foam or an elastomer, with small amounts of a filler. Doping these materials (which have a permeability of 1) with a high dielectric loss material such as carbon or graphite modifies the dielectric properties of the materials. In contrast, with magnetically loaded materials, fillers such as ferrites, iron, or cobalt nickel alloys increase the permeability of the base material. By doing so, they create a material that is still essentially non-conductive, but has "chains" of conductive molecules that create the loss or absorption as the wave passes through it. These chains are big enough to attenuate the signal so it cannot pass though unhindered, but are also small enough that they cannot sustain a sufficient "surface" current; therefore they do not reflect any energy.

The most commonly used absorbers tend to be magnetically doped elastomers. Foam equivalents can be made from either dielectric or magnetic materials, but are less widespread due to their size. That is, they tend to be thicker to achieve the desired absorption.

So what makes absorbers useful? Let's paint a scenario. You've designed your system doing all you can do in the electronics for suppression. You have designed a spectacular shield that isolates your radio from the remaining RFI. So you thought. It turns out that despite your best efforts, the enclosure design, including the shield, has become what is known as a *shielded cavity,* which at certain frequencies, based on the cavity geometries, will act like a resonant cavity. At these resonant frequencies, very small amounts of noise can easily be amplified by $> 10\times$ (20 dB), creating sufficient energy within your platform to cause functional/performance issues not only in the radios but also in other system functional units and digital communication buses. Absorbers offer a possible solution to this issue.

Absorbers are actually nothing new having been used in the military for some time now. Absorbers can be found throughout the defense industry from ship antenna arrays to prevent false radar images to use on stealth aircraft where the absorber is used to prevent radar detection. In recent years, with a continuing trend toward higher and higher clock frequencies and the widespread use of wireless devices, there is now a considerable market in the commercial space for absorbers. Examples of applications range from notebook computers to network servers and switches to wireless antenna systems, including cellular base stations, not to mention wireless LAN devices. In much of these applications, absorbers are used as a simple shielding solution or to dampen these cavity resonances.

With the increasing clock and processor operating frequencies, conventional shielding solutions become more complex and difficult to implement. Therefore, more and more manufacturers are turning to absorbers for a solution. Wireless proliferation is a major driver in this trend, with the higher operational frequencies creating greater interference (hence the timing of this publication). Microwave absorbers, are they are also known, can address frequencies from 500 MHz to multi GHz. Figure 8.15 shows a selection of performance curves for some absorbers.

As we can see in Figure 8.15, absorbers tend to have what can be called a *center* or *tuned* frequency of operation. Absorption loss, also known as *reflection attenuation,* decreases either side of this center frequency. Herein is the biggest disadvantage of absorbers. If you need to attenuate a wide range of frequencies, then this becomes a challenge. Additional characteristics of commercially available absorbers are shown in Figure 8.16.

We can also see the other disadvantages of absorbers in Figure 8.16. They need to be pretty thick to meet performance requirements in the low GHz, and they weigh a lot. If we need

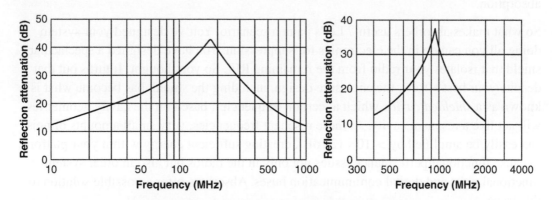

Figure 8.15: Performance curves of absorbers.

Center Frequency (GHz)	20 dB Bandwidth (GHz)	Standard Thickness (mm)	300 300 mm standard weight (kg)	Application
0.9	0.08	8.4	2.4	Cellular Phone
1.5	0.15	6.8	1.8	False Echoes (Radar)
2.4	0.35	5.9	1.4	Wireless LAN
5.1	1.10	5.3	1.2	Wireless LAN

Figure 8.16: Examples of absorber characteristics.

to absorb 10 GHz, then absorbers may be a good choice. In wireless bands, there still remain many challenges to make them work. Saying this, some manufacturers do use them and see them as a viable solution. This is due partly to the fact that due to time constraints, circuit-board and enclosure designers alike generally do not go through the complex resonance modeling exercise necessary to identify where and what frequency resonances may exist in their design. Absorbers are therefore seen as a last-minute savior, with system designers employing the practice of trying different absorber samples and cutting and pasting different materials until the problem is solved.

To illustrate this, we'll look at some data. Given that antennas tend to be placed in or adjacent to LCD panels, we will study the specific impacts of absorbers in such an application.

We can see in Figure 8.17 that an absorber was placed adjacent to the WLAN antenna on the top edge of the LCD. Using a 1-inch-by-3-inch piece of absorber placed between the antenna and the LCD panel, the following data was measured.

As shown in Figure 8.18, the absorber performed fairly well with as much as a 20 dB improvement in the upper-frequency range (> 1,700 MHz), but not so much in the lower-frequency range (700 MHz to 1,000 MHz). Some improvement was observed at the lower frequencies, but the benefit was not consistent. Using the same approach, testing was carried out on a second platform from a different manufacturer. The results are shown in Figure 8.19.

As shown in this example, the absorber helped in both the lower- and upper-frequency bands. Given that the same absorber was used for both platforms, the data reveals one of

Figure 8.17: Absorbers in LCD panel applications.

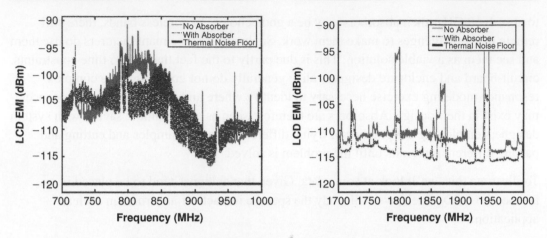

Figure 8.18: Absorber improvements (platform #1).

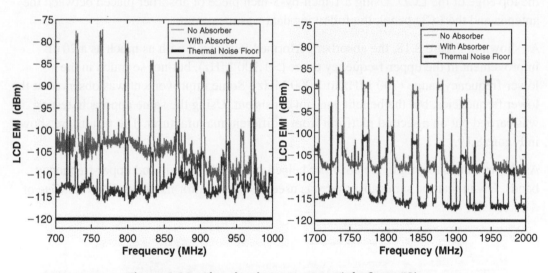

Figure 8.19: Absorber improvements (platform #2).

the significant drawbacks of absorbers. What works in one platform may not work in another! Cost is another important factor here. Each 1-inch-by-3-inch piece of absorber used here was priced between $1.50 and $2.25 (in volume). If we assume that an equivalent piece of absorber is needed for every antenna on the platform, that would equate to an added cost of anywhere between $3 for a single radio (dual antenna) system to $15 for the multi-radio platforms somewhere in the not too distant future. This is clearly not appropriate in the consumer space and not something that will gain you friends in high places. Taking this into account as well as the weight and size impacts, the value proposition for absorbers may be a tough sell.

Fortunately, the world does not stand still. Absorbers are getting better, due in part to demand. With the emergence of ever smaller form factors and the customer demand for mobility, making wireless functionality a must, today's target applications need thin and light absorbers.

In the meantime, if you think you may need an absorber, plan the space for it. Retroactive absorber fitting is a very painful task—one that will not gain you any respect with either your design colleagues or your management.

8.2.3 Layout

Layout is one of the most powerful tools at the hands of the designer. By using good layout techniques, sensitive radios can be placed away from noisy components. It may seem overly simple, but distance can be your friend when it comes to noise. To explain this, we'll first look at the theory. In Chapter 3 when we addressed measurements, the concept of near-field to far-field transition was introduced. In the shielding section earlier in this chapter, we also talked about near-field versus far-field performance of shields. Again, we will start by looking at the far field.

The respective magnetic and electric fields observed at a distance r from a current loop can be described by

$$H_{A/M} = \frac{\pi I A}{\lambda^2 r}$$

$$E_{V/m} = \frac{120 \pi^2 I A}{\lambda^2 r},$$

where

I = Loop Current (Amps)
A = Loop Area (square meters)
λ = Wavelength (meters)
r = Distance (meters).

We can see from these equations that both the magnetic (H) and electric (E) field fall off as the inverse of the distance, that is, $1/r$. In simple terms, every time we double the distance between the observation point and the source of the noise, we see a $2\times$ or 6 dB reduction.

From these equations, we can also see that the ratio of electric to magnetic field (known as the *wave impedance*) is constant here. That is

$$E/H = 120\pi = 377 \, (\text{ohms}).$$

We have already stated that for small form factor devices it is more than likely we will have victim devices such as radios in the near field. The equivalent expressions for the near field are

$$H_{A/m} = \frac{IA}{4\pi r^3}$$

$$E_{V/m} = \frac{60\pi IA}{\lambda r^2}.$$

In the near field, we see a significant difference in behavior. The H field is falling off as $1/r^3$, and the electric field is falling off as $1/r^2$. This is noteworthy as it tells us that any movement either closer or further away in the near field will see a severe change in the observed field strengths. So, in this case, moving $2\times$ further away decreases the E field by $4\times$ (12 dB) and the H field by $8\times$ (18 dB).

It is therefore clear that moving the victim further away from a potential noise source is a powerful tool in our arsenal of potential solutions.

These expressions only deal with radiated noise and do not represent all potential coupling paths such as conducted noise, but if you have established radiated coupling as a primary coupling mechanism, then layout can obviously help significantly. We will now look at a real-life example of this.

8.2.3.1 Antenna Placement

The majority of laptop manufacturers place their antennas in the LCD panel today. They do this for a variety of reasons, including better coverage, omni-directionality, as well as a belief that there is lower noise in the LCD clamshell. The antennas tend to be placed along the edges of the panel and primarily along the top edge for obvious reasons. We'll now see if location along these edges can play a role in the level of noise seen by the antenna.

Using the near-field techniques described in Chapter 5, a noise capture at 2.41 GHz as shown in Figure 8.20 was measured. For information purposes, we have annotated the row (source) and column (gate) drivers associated with the panel functionality.

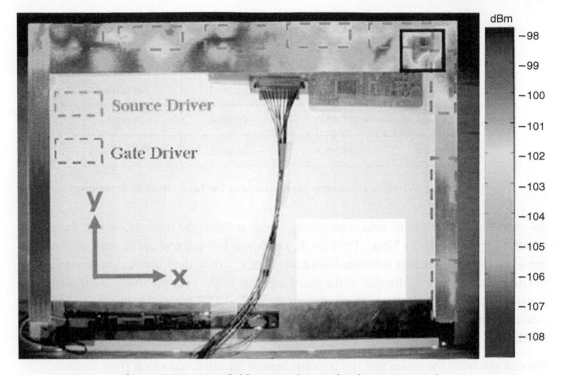

Figure 8.20: Near-field scans of a production LCD panel.

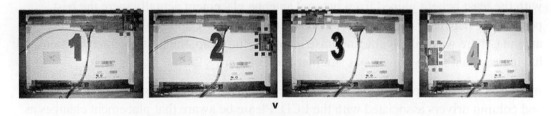

Figure 8.21: Targeted antenna positions.

The first observation is that the top-right corner is particularly "hot" as is the top edge compared to any other part of the panel. Armed with this information, it would seem clear that there are good and bad positions for the antennas. The logical next step is to see if these results correlate with antenna measurements.

Using a WLAN antenna at the four positions shown in Figure 8.21, sensitivity levels were measured in three WiFi channels. This is shown in Figure 8.22.

Channel Center Frequency	2410 MHz	2442 MHz	2475 MHz
Position #1	−92 dBm	−93.5 dBm	−96.75 dBm
Position #2	−108 dBm	−110 dBm	−106 dBm
Position #3	−103 dBm	−105 dBm	−109 dBm
Position #4	−110 dBm	−108 dBm	−109 dBm

Figure 8.22: Wireless sensitivity measurements for four antenna locations.

As expected from the near-field scan results, position 1 was the noisiest, and all three other positions were markedly better. Position 4, which was the quietest in the scan, also turned out to be the best antenna location. Using this data, it would therefore be prudent to position the antenna on the side of the panel adjacent to the column divers. To illustrate the possible benefits of antenna placement, Figures 8.23 to 8.27 highlight the maximum and minimum noise levels for similarly positioned antennas for different wireless radio bands.

We can see from the tables that the possible benefit for noise reduction from antenna placement can range from 0 dB to as much as 13 dB. It is interesting to note that there is a wide variety of performance from the five panel manufacturer samples, and that the panels with the least benefit gained from optimal antenna placement actually were the best performing panels of the samples. Also note that even after optimal placement of the antenna, there was still significant noise impact to the radio, the best antenna positions still having residual noise impacts, as summarized in Figure 8.28.

In all cases, the best performance was realized by placing the antenna away from the row and column drivers associated with the LCD. Please be aware that placement changes as shown above will also affect the performance of the antenna in terms of coverage, directionality, and so on. It is therefore recommended that any "quiet" antenna positions be tested for antenna performance to ensure that the appropriate tradeoffs are made between lowest noise and best antenna performance. It is also clear from this study that there are limited "quiet" areas. If your design requires multiple antennas due to either the use of MIMO (Multiple Input Multiple Output) radios or indeed multiple radios (i.e., WWAN and WLAN), it will be difficult to find quiet locations for all of them. Given that current trends point to future designs having up to four different radios and perhaps seven different antennas, placement alone is unlikely to solve LCD-generated noise problems.

MFR	Peak Desense (dB)	Min Desense (dB)	Antenna Placement Benefits (dB)
Manu 1 15"	30.3	15.06	15.24
Manu 2 14.1" XGA	21.07	10.5	10.57
Manu 3 14.1" XGA	28.8	12.5	16.3
Manu 4 14.1" XGA	11.36	10.38	0.98
Manu 5 15.0"	19.37	10.75	8.62

Figure 8.23: WWAN, 869–894 MHz, Channel BW = 200 KHz.

MFR	Peak Desense (dB)	Min Desense (dB)	Antenna Placement Benefits (dB)
Manu 1 15.0"	24.96	16.29	8.67
Manu 2 14.1" XGA	16.82	8.51	8.31
Manu 3 14.1" XGA	22.35	14.25	8.1
Manu 4 14.1" XGA	7.1	6.9	0.2
Manu 5 15.0"	18.08	13.2	4.88

Figure 8.24: WWAN, 925–960 MHz, Channel BW = 200 KHz.

MFR	Peak Desense (dB)	Min Desense (dB)	Antenna Placement Benefits (dB)
Manu 1 15"	23.85	11.89	11.96
Manu 2 14.1" XGA	21.38	10.71	10.67
Manu 3 14.1 XGA	20.87	12.22	8.65
Manu 4 14.1" XGA	13.95	6.31	7.64
15.0" Manu 5	26.6	13.2	13.4

Figure 8.25: WWAN, 1805–1880 MHz, Channel BW = 200 KHz.

MFR	Peak Desense (dB)	Min Desense (dB)	Antenna Placement Benefits (dB)
Manu 1 15″	20.06	8.22	11.84
Manu 2 14.1″ XGA	18.1	6.95	11.15
Manu 3 14.1″ XGA	17.03	12.87	4.16
Manu 4 14.1″ XGA	12.41	6.69	5.72
Manu 5 15.0″	13.92	8.2	5.72

Figure 8.26: WWAN, 1930–1990 MHz, Channel BW = 200 KHz.

MFR	Peak Desense (dB)	Min Desense (dB)	Antenna Placement Benefits
Manu 1 15″	8.94	4.09	4.85
Manu 2 14.1″ XGA	4.6	4.25	0.35
Manu 3 14.1″ XGA	6.1	4.37	1.73
Manu 4 14.1″ XGA	6.51	4.18	2.33
Manu 5 15.0″	6.38	4.54	1.84

Figure 8.27: WLAN, 2400–2500 MHz, Channel BW = 20 MHz.

Radio Band	WWAN (800 MHz)	WWAN (900 MHz)	WWAN (1800 MHz)	WWAN (1900 MHz)	WLAN (2.4,2.5 GHz)
Max Noise Impact	30.3 dB	24.96 dB	26.6 dB	20.06 dB	8.94 dB
Min Noise Impact	10.38 dB	6.9 dB	6.31 dB	6.69 dB	4.09 dB
Placement Benefit Range	0.98–15.24 dB	0.2–8.67 dB	7.64–11.96 dB	5.72–11.84 dB	0.35–4.85 dB

Figure 8.28: Summary of antenna placement results.

In the above study, we were limited to moving the location of the victim, in this case the antenna itself. There is of course the alternate strategy of making sure the source of the noise is as far away from the victim device as possible. This entails having a deeper understanding of potential noise sources within your device or system and some degree of control in the design process. It is a well-known EMI/EMC design practice to keep high-frequency and low-frequency circuits or components apart. By doing so, the EMC engineer limits coupling between analog and digital functions, for example, allowing the engineer to limit the more severe design techniques to high-risk/high-frequency areas, while applying only minimal techniques to the low-risk/low-frequency areas. This only works, however, if no coupling between the two areas can be guaranteed. In some products, this is done by physically positioning the "similar" functional blocks together and then creating shields between them. We can see examples of this in Figures 8.29 and 8.30.

In Figure 8.29, we can see three distinct areas: two "shielded" areas where three components have been placed together in one, and a single component in the other. There is one additional area outside of the shield where an IC and some passives are placed. By placing the components appropriately and then creating the shield, the designer here is attempting to create isolation between these three distinct sections or functional blocks.

In Figure 8.30, there are significantly more divided sections. In cases like this, a designer sometimes has the added flexibility of separating particularly noisy areas from sensitive sections and thus creating intermediate areas to increase the achievable isolation. There is one word of warning. By placing features or components in between the noisy and sensitive sections, you may end up with less isolation than expected as the noise can be

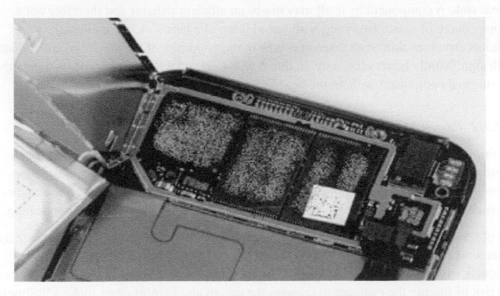

Figure 8.29: Placement/shielding example #1.

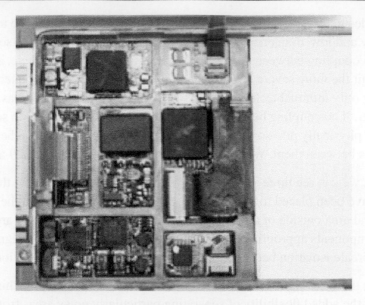

Figure 8.30: Placement/shielding example #2.

coupled and re-radiated. This typically happens if the structure or component is a more efficient radiator based on geometry than the original source. An example of this would be a heat sink. A component by itself may not be an efficient radiator and therefore not a strong source of noise. However, with the addition of a heat sink, a significantly more efficient structure akin to an antenna is now in the proximity of the noise and can radiate with significantly better efficiency. In this way, we end up with a higher radiated field than the original component without the heat sink.

8.2.3.2 Passive Mitigation Roundup

So far, we have covered the potential benefits of shielding: absorbers and layout. Each approach can be very useful and can be used in combination to improve/gain isolation between noisy platform components and victim radios. We discussed the relative merits for each approach, as well as the potential pitfalls of near-field and far-field application.

All of these approaches suffer from a similar fundamental flaw. There is no guarantee that what worked in one product will work in another. With different geometries, different products will need different solutions. Even the slightest of changes in the form factor run the risk of forcing the engineer to reassess the design and in most cases make adjustments.

These approaches therefore force a build-it-and-see design process, which runs the risk of last-minute issues, added product cost, and delayed product release.

In the next chapter, we will address what we call *active mitigation*. Unlike passive mitigation, which assumes that we have a noise source and sensitive victim receiver and attempts to maximize the isolation between them, active mitigation deals with the source of the noise itself.

References

[1] C. Paul, *Introduction to Electromagnetic Compatibility* (2nd ed.), Wiley, 2006.

[2] H.W. Ott, *Noise Reduction Techniques in Electronic Systems* (2nd ed.), Wiley, 1988.

[3] M. Mardiguian, *Controlling Radiated Emissions by Design,* Springer, 2001.

[4] D.R.J. White, *Electromagnetic Interference and Compatibility,* vol. 3, *EMI Control Methods and Techniques*, Don White Consultants, 1973.

[5] A.F. Molisch, *Wireless Communications,* Wiley, 2005.

[6] TDK Application Note: *EMC Components Radio Wave Absorbers: Application Examples of Electromagnetic Absorbers* (Ref 001-03/20070302/e9e_bdj_003). *EMI/RFI Shielding Frequently Asked Questions,* Thermospray Company,

[7] W.D. Kimmel, and D.D. Gerke, *EMC Design for Compliance: Conductive Coatings* (Conformity Whitepaper).

[8] A. Sundsmo, *Microwave Absorbers: Reducing Cavity Resonances,* Laird Technologies (Compliance Engineering Whitepaper).

[9] J. Armstrong and H.A. Suraweera, *Impulse Noise Mitigation for OFDM Using Decision Directed Noise Estimation.* ISSSTA2004, Sydney, Australia (30 Aug to 2 Sept 2004).

[10] T. Yucek and H. Arslan, *MMSE Noise Power and SNR Estimation for OFDM Systems.* (Whitepaper, University of South Florida).

[11] A.N. Barreto and S. Furrer, *Adaptive Bit Loading for Wireless OFDM Systems.* IBM Zurich Research Laboratory, Ruschlikon, Switzerland.

[12] K.M. Nasr, F. Costen, and S.K. Barton, *Performance of Different Interpolation Strategies for OFDM/MMSE Smart Antenna System in an Indoor WLAN.* The University of Manchester, UK. Vehicular Technology Conference, 2005.

[13] Y. Kakishima, H. Le, S.H. Ting, K. Sakaguchi, and K. Araki, *Experimental Analysis of MIMO-OFDM Eigenmode Transmission with MMSE Interference Canceller.* Tokyo Institute of Technology. The 17th Annual IEEE International Symposium on Personal, Indoor and Mobile Radio Communication (PIMRC 2006).

CHAPTER 9

Active Mitigation

In this chapter, we will discuss the following topics: frequency planning, frequency content, and radio improvements.

9.1 Frequency Planning

In Chapter 8, we briefly discussed the concept of using layout as a mechanism to keep high-frequency/high-noise components away from our sensitive radio components. In this chapter, we'll take the same fundamental concept but apply it to frequency rather than distance. Taking the concept to its extreme, the idea is to avoid operational frequencies that coincide with the targeted radios for your system, and thereby completely avoid the problem of interference in the first place. What makes it tough is that we also need to make sure that the harmonics of the operational frequencies do not overlap with our radios. This is illustrated in Figure 9.1.

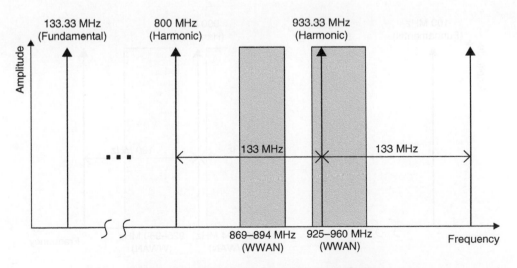

Figure 9.1: Frequency planning example #1.

Plotted in Figure 9.1 are two radio bands in the 800/900 MHz range and a 133 MHz noise source and its harmonics. In order to simplify the figure, we have only included the fundamental and the harmonics of interest. As you can see, the 7th harmonic sits in the 925–960 MHz WWAN band. In this current configuration, there is a distinct possibility that the noise at 933 MHz may interfere with the radio operating in the same range.

If rather than using a 133 MHz operational frequency we had selected 100 MHz, the picture would be quite different. As you can see in Figure 9.2, we no longer have any overlap, and any noise generated by the 100 MHz fundamental or its harmonics is a don't care from an RFI perspective.

Given this scenario, the designer would not have to worry about shielding, absorbers, or any after-the-fact solutions, since there would be no energy or platform-generated noise in the radio band. This of course is an oversimplification, given that there will be multiple operational frequencies and therefore a multitude of harmonics to deal with. Nevertheless, it can be applied at a minimum as an approach to minimize the number of noise sources and thereby reduce the focus of mitigation efforts to a manageable few.

In Figure 9.2, we changed the fundamental frequency from 133 MHz to 100 MHz. This is quite a large change and in all likelihood would not be acceptable, as it would adversely affect the performance of the system. In reality, the required change to move the noise out of the radio band need not be so drastic. We'll look again at our example.

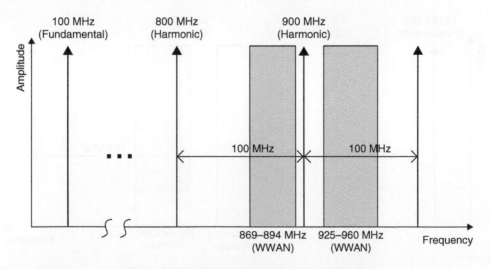

Figure 9.2: Frequency planning example #2.

In the original case in Figure 9.1, with the operational frequency of 133.33 MHz, the 7th harmonic fell into our WWAN GSM band at 933.33 MHz. To move this out of the band, we would only need to move it by approximately 9 MHz. This takes it below the 925 MHz start frequency for our radio band. Given that this is the 7th harmonic, this means that we only need to move the fundamental frequency by 9/7 MHz = 1.3 MHz. To move the harmonic out of the target radio band, we therefore only need to move the fundamental frequency from 133.33 MHz to 132 MHz, a reduction of just under 1%. In this way, we can move the frequency while not impacting the performance of the system, a much more palatable solution. Although in our example we reduced the fundamental frequency to move the harmonic out of the lower edge of the radio band, it clearly follows that if the harmonic is actually closer to the upper band edge, then the fundamental frequency can be increased to move the harmonic up rather than down in frequency. This concept is depicted in Figure 9.3.

Looking at the figure, we see the original frequency f_o, which results in a harmonic collision in a radio band at frequency nf_o. In order to move the harmonic out of the radio band, we need to move it by 'X' MHz as shown in the figure. Given that the collision happens at harmonic n, we can state that

$$X = n\left(f_o' - f_o\right).$$

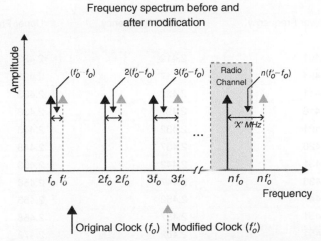

www.newnespress.com

Figure 9.3: Frequency planning; Tweaking operational frequencies to move noise out of band (OOB).

From this, we can derive that the required new fundamental frequency can be expressed as

$$f_o' = \frac{nf_o + X}{n}$$
$$= f_o + \frac{X}{n}.$$

It is therefore a simple matter to calculate what the new fundamental frequency should be to move the harmonic noise of the problem radio channel. If we need to move the frequency down rather than up, X becomes negative in this expression resulting in an f_o' lower than the original f_o.

This approach of course assumes that the designer has sufficient flexibility and granularity to move the fundamental frequency where required. Sometimes this may not be the case, and only coarse adjustments may be possible. We'll take the example of a 65 MHz pixel clock causing interference in the WLAN RF band. Doing a quick calculation, we see that

$$65\,MHz^*37\,(37th\,Harmonic) = 2405\,MHz$$
$$65\,MHz^*38\,(38th\,Harmonic) = 2470\,MHz.$$

The table in Figure 9.4 shows U.S. WLAN channels. We can see that Channels 1, 2, and 11 may be impacted by noise at these frequencies.

Channel	Lower Frequency	Center Frequency	Upper Frequency
1	2.401	2.412	2.423
2	2.404	2.417	2.428
3	2.411	2.422	2.433
4	2.416	2.427	2.438
5	2.421	2.432	2.443
6	2.426	2.437	2.448
7	2.431	2.442	2.453
8	2.436	2.447	2.458
9	2.441	2.452	2.463
10	2.451	2.457	2.468
11	2.451	2.462	2.473

Figure 9.4: U.S. WLAN channel definition.

Given that the total WLAN band is ~ 75 MHz (in the U.S.), it is clear that we cannot move the noise completely out of all channels simultaneously unless we move the fundamental frequency to something greater than 75 MHz. Doing a quick calculation as shown in Figure 9.5 reveals that if you are limited to 1 MHz granularity, then you must change the frequency to 92 MHz in order to avoid any potential collisions entirely. In this table, collisions are indicated by a highlighted cell. The highlighted line that includes the frequency indicates no collisions. 75 MHz itself is a little too close for comfort sitting right on the boundaries for channels 1 and 11. Giving yourself around 500 KHz to 1 MHz margin, either side of the radio channel is a good idea.

Fundamental Frequency	Harmonic Orders											
	25	26	27	28	29	30	31	32	33	34	35	36
74						2220	2294	2368	2442	2516	2590	
75						2250	2325	2400	2475	2550		
76						2280	2356	2432	2508			
77						2310	2387	2464	2541			
78						2340	2418	2496				
79						2370	2449	2528				
80					2320	2400	2480					
81					2349	2430	2511					
82					2378	2460	2542					
83				2324	2407	2490						
84				2352	2436	2520						
85				2380	2465	2550						
86			2322	2408	2494							
87			2349	2436	2523							
88			2376	2464	2552							
89		2314	2403	2492								
90		2340	2430	2520								
91		2366	2457	2548								
92		2392	2484	2576								
93	2325	2418	2511	2604								
94	2350	2444	2538	2632								

Figure 9.5: Spreadsheet calculation of harmonic collisions (1 MHz granularity).

If you have finer granularity than 1 MHz, perhaps 0.1 MHz, then a solution exists at a much lower frequency. This is shown in Figure 9.6. Here you can see that solutions exist at both 77.4 MHz and 79.9 MHz.

This, however, may still be problematic as they represent significant percentage changes in frequency. There is one additional facet we can consider before closing this topic. If we assume that we only have 1 MHz granularity as in the first case, what can we do, while keeping the fundamental frequency within a couple of percentage points?

Fundamental Frequency	Harmonic Orders							
	29	30	31	32	33	34	35	36
75		2250	2325	2400	2475	2550	2625	
76		2280	2356	2432	2508	2584	2660	
77		2310	2387	2464	2541	2618	2695	
77.1		2313	2390.1	2467.2	2544.3	2621.4	2698.5	
77.2		2316	2393.2	2470.4	2547.6			
77.3		2319	2396.3	2473.6	2550.9			
77.4		2322	2399.4	2476.8	2554.2			
77.5		2325	2402.5	2480	2557.5			
78		2340	2418	2496				
79		2370	2449	2528				
79.1		2373	2452.1	2531.2				
79.2		2376	2455.2	2534.4				
79.3		2379	2458.3	2537.6				
79.4		2382	2461.4	2540.8				
79.5		2385	2464.5	2544				
79.6		2388	2467.6	2547.2				
79.7		2391	2470.7	2550.4				
79.8		2394	2473.8	2553.6				
79.9		2397	2476.9	2556.8				
80	2320	2400	2480	2560				

Figure 9.6: Spreadsheet calculation of harmonic collisions (0.1 MHz granularity).

Fundamental Frequency	Harmonic Order							
	29	35	36	37	38	39	40	36
63		2205	2268	2331	2394	2457	2520	
64		2240	2304	2368	2432	2496	2560	
65		2275	2340	2405	2470	2535	2600	
66		2310	2376	2442	2508	2574	2640	
67		2345	2412	2479	2546			

Figure 9.7: Spreadsheet calculation of harmonic collisions (even vs. odd harmonics).

Looking at Figure 9.7, we see that with our original frequency of 65 MHz, we ended up with two harmonics in the WLAN band. If we look either side of 65 MHz, we'll see what our options are.

The good news is that all of the alternatives have the benefit of only having one harmonic in the band. Looking more closely, 64 MHz and 67 MHz have one other benefit. So far in our analysis, we have essentially assumed that all harmonics have the same amplitude and therefore the potential performance impacts to radios are equivalent. Fortunately, this is not the case. As mentioned in the analysis in Chapter 2, duty cycle is one of several factors that have a direct impact on spectral content. The concept of using spectral manipulation is covered in more detail in the next section, but as somewhat of an introduction, even and odd harmonics may have vastly different spectral signatures. In actual fact, if a signal has a perfect 50% duty cycle, in theory it will have no even harmonics. This gives us an additional weapon in our arsenal. In our example above, by moving from a 65 MHz fundamental to either 64 MHz or 67 MHz, we move the offending noise from being an odd to an even harmonic. We can't expect there to be no energy at the even harmonic, as real designs don't have perfect 50% duty cycle attributes, nevertheless it has the opportunity to be significantly less. Having touched on the topic of frequency content, we'll now move on and address this in detail.

9.2 Frequency Content

One of our foundational concepts throughout this publication has been the idea that unlike EMI, for platform interference in wireless systems we can trade off noise inside and outside of the radio bands. In the previous section, we achieved this by moving the noise out of radio bands and into frequencies where no radio operates. In this section, we will address the fundamental aspects of digital circuitry that impact frequency content, and from that we will derive possible mitigation approaches.

We'll start by looking at typical digital signals and their associated frequency content. Figure 9.8 shows a representation of a trapezoidal pulse.

In Figure 9.8, we have a 50% duty cycle with identical rise (t_r) and fall (t_f) times. The period $T = 1/f$ and $\tau = T/2$ are also annotated. The Fourier envelope for such a pulse train is shown in Figure 9.9.

As we can see, there are two break points. The first break point, F_1, is a function of the period; the second break point, F_2, is a function of the rise (or fall) time. Below the first break point, the spectral content is flat. Between the first and second break points, the

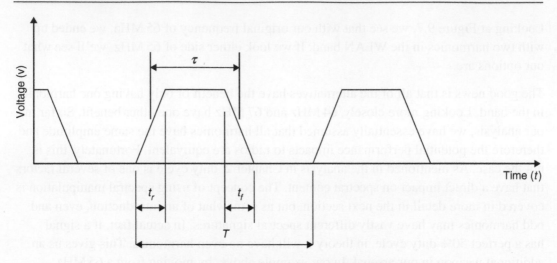

Figure 9.8: 50% duty cycle trapezoidal signal.

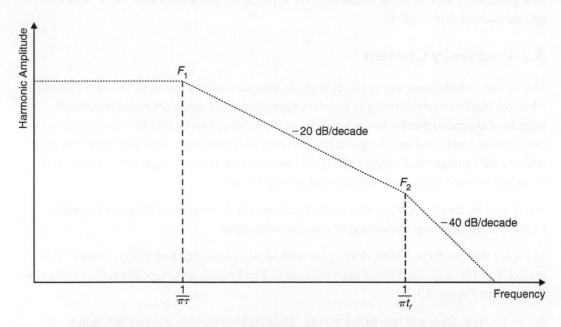

Figure 9.9: Fourier envelope for a 50% duty cycle trapezoidal waveform.

harmonic energy will decrease at −20 dB/decade. After the second break point, the harmonic amplitude will decrease at −40 dB/decade. We'll use an example to illustrate the importance of this characteristic.

We'll assume a 100 MHz pulse train with a perfect 50% duty cycle and 1 ns rise and fall times. The respective values are therefore T = 10 ns, τ = 5 ns, and $t_r = t_f = 1$ ns.

The first break point $(F_1) = \dfrac{1}{\pi\tau} = 63.67$ MHz.

The second break point $(F_2) = \dfrac{1}{\pi t_r} = 318.3$ MHz.

This means that we should expect a −20 dB/decade fall off of harmonic amplitude between 100 MHz and 300 MHz, and a −40 dB/decade fall off for all higher-order harmonics above this. Now we'll change the rise time and fall time to 200 ps and see what happens.

The first break point (F_1) remains unchanged $= \dfrac{1}{\pi\tau} = 63.67$ MHz.

The second break point (F_2) now $= \dfrac{1}{\pi t_r} = 1.59$ GHz.

With the second break point now at ∼1.6 GHz, there will be significantly more harmonic amplitude for all harmonics between 400 MHz and 1.6 GHz.

If we do a comparative analysis, for example 900 MHz, we will see the following possible difference in amplitude due to the rise time change.

For a 1 V trapezoidal waveform:

$$\text{For Frequencies } F \text{ where } \quad F_1 < F < F_2, \quad \text{Amplitude} = \frac{2}{\pi\tau F}$$

$$(F \text{ term equates to} - 20\,\text{dB/decade slope})$$

$$\text{For Frequencies } F \text{ where } \quad F_2 < F, \quad \text{Amplitude} = \frac{0.2}{2\tau t_r F^2}$$

$$(F^2 \text{ term equates to} - 40\,\text{dB/decade slope})$$

With a 1 ns rise time, 900 MHz is beyond the second break point F_2, so

$$\text{Amplitude} = \frac{0.2}{2 \times 5 \times 10^{-9} \times 1 \times 10^{-9} \times \left(900 \times 10^6\right)^2} = 0.024\,\text{V} = 87.6\,\text{dB}\mu\text{V}.$$

With a 200 ps rise time, 900 MHz was between F_1 and F_2, so

$$\text{Amplitude} = \frac{2}{\pi \times 5 \times 10^{-9} \times 900 \times 10^6} = 0.141\,\text{V} = 103\,\text{dB}\mu\text{V}.$$

The absolute values are not important here and will change based on a variety of factors. The difference will remain constant, and we can see that by simply changing the rise time from 1 ns to 200 ps, we increased the harmonic amplitude by more than 15 dB.

It is an unfortunate fact that in today's digital systems, the designer (electrical timing) will invariably make the rise and fall times faster than is really necessary. The authors of this book have seen 14 MHz clocks with rise times less than 100 ps, when functional requirements would have been met by a 7 ns rise time (a 70× difference). When a 5× difference can lead to a 15 dB increase in amplitude at a higher-order harmonic, imagine the difference with a 70×-faster rise time than is really necessary. With such rise times, in fact, 14 MHz harmonics have been measured out past the 120th harmonic. This is illustrated in Figure 9.10.

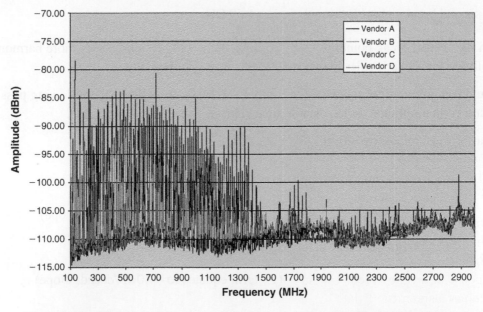

Figure 9.10: GTEM comparative measurements of functionally identical parts.

Shown here is a series of plots from four suppliers of a system timing component. Although all of these parts are functionally identical, they have vastly different spectral content with up to 20 dB difference in noise levels. This is a prime example of a component specification that focuses on edge rates that are "faster than X" and stipulates no upper bound. In this way, equivalently specified parts can have dramatically different spectral performance. The lesson here is to make sure you specify maximum and minimum values for edge rates in your designs and associated component specification.

9.2.1 Spread Spectrum Clocking

We covered spread spectrum clocking briefly in Chapter 1. Here we look at the topic in greater detail with a focus on how it may be used as a mitigation technology.

One of the most common spectral control mechanisms used in the personal computer and consumer electronics industries is something called *clock dithering* or *Spread Spectrum Clocking (SSC)*. The basic concept is that periodic timing signals such as clocks or strobes are frequency modulated to "spread" their typically narrowband energy over a wider frequency range. For EMI, this lowers the instantaneous electromagnetic energy in a given frequency range, thus in theory lowering the interference potential of the potential aggressor. Regulatory EMI measurements can benefit by anywhere between 6 and 20 dB by using this approach.

There is much debate in the RF community on whether SSC does actually lower the interference potential. It is not the goal of this publication to weigh in on this debate. To see how SSC may or may not help us, we'll first look at the fundamentals.

To understand how SSC may impact a radio, a fundamental understanding of both the time and frequency domain is required. In the frequency domain, the effect of SSC can be seen in Figure 9.11.

Shown here is a non-spread 97 MHz clock and its 0.6% "center-spread" equivalent. The term *center spread* indicates that the modulation is centered on the original non-spread frequency with excursions both higher and lower in frequency (+0.3/−0.3%). Due to the response of the measurement equipment, it looks like the SSC signal continuously occupies a significantly larger spectrum than a non-SSC signal. As discussed in Chapter 1, with this observation it is understandable why certain parties would believe that although the level may be decreased, the interference potential would be higher due to the continuous occupation of a larger spectrum (MHz vs. KHz). As we discussed earlier, when we consider the time domain response of spread, we can reach a much different conclusion.

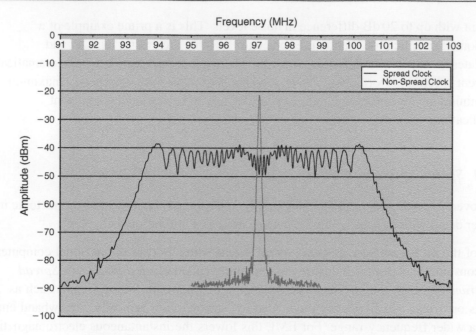

Figure 9.11: Frequency domain plot of clock dithering.

This is because the signal does not continuously occupy the spectrum as indicated on a spectrum analyzer or similar device, but only occupies a bandwidth similar to a non-SSC signal at any single snapshot in time. At an incremental snapshot in time later, the SSC signal will occupy a similar BW but at a different center frequency. By dithering the clock, the frequency of the fundamental and its harmonics vary with time over a total spectrum consistent with the spread percentage. The amount of time spent at any one frequency will depend on the SSC implementation, that is, the modulation signature (linear, sinusoidal, non-linear) and frequency (KHz). Figure 9.12 shows examples of linear and "Hershey's Kiss" (non-linear) type modulations for a 100 MHz 0.5% down spread clock.

For wireless devices, the interference potential of the signal is dependent on the presence of the aggressor in a radio band for a length of time significant with respect to the integration time of the receiver. It is easy to understand why stationary interference signals can impact a radio. To what extent our moving "spread" clock impacts the wireless radio is somewhat more complex.

We will consider two cases: when the spread signal is wholly contained within the radio channel, and when the spread signal is partially inside and outside the radio channel.

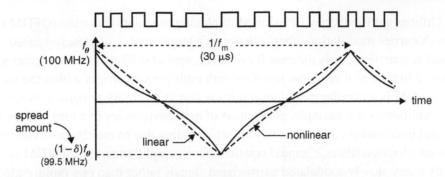

Figure 9.12: Linear and "non-linear" SSC modulation profiles.

We'll first show an example of when the spread signal is wholly contained within the radio channel. Figure 9.13 shows a 0.6% spread clock with a harmonic falling in the 802.11b/g 20 MHz wide channel at 2.4 GHz. Here the spread signature is approximately 14 MHz wide and is fully contained within the channel of interest.

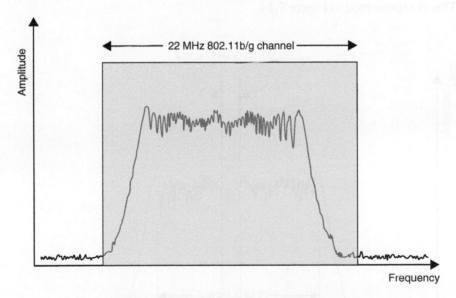

Figure 9.13: Spread clock in a radio band.

In this example, spreading the clock is unlikely to help the radio at all. All of the energy/power is still contained within the radio channel. It actually has the potential to be worse than a narrowband signal, dependant on the type of radio receiver and modulation used. If Orthogonal Frequency Division Multiplexing modulation is used (OFDM is a digital multicarrier modulation scheme that uses a large number of closely-spaced orthogonal subcarriers), as is the case for 802.11g, spread is likely to have a greater performance impact as it will now interfere with multiple subcarriers within the radio channel. OFDM is a popular modulation scheme due to its ability to cope with severe channel conditions—for example, attenuation of high frequencies in a long copper wire, narrowband interference, and frequency-selective fading due to multipath—without complex equalization filters. Channel equalization is simplified because OFDM is essentially many slowly-modulated narrowband signals rather than one rapidly-modulated wideband signal. Because OFDM can more easily cope with narrowband interferes, spread in this case with all the energy contained within the channel does not help, and we lose some of the potential benefits of OFDM.

In the next example, we'll show when the spread signal is partially inside and outside the radio channel. Here we use the example of a similar 0.6% spread clock in the 1900 GSM band, where the effective radio bandwidth is 200 KHz and the spread signature is 11 MHz wide. This is represented in Figure 9.14.

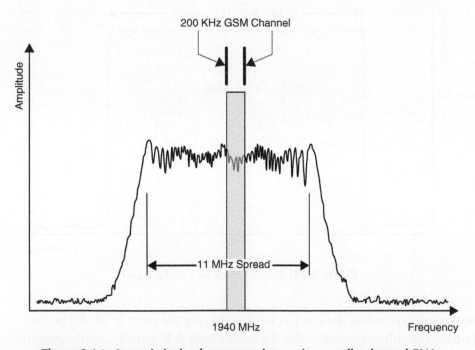

Figure 9.14: Spread clock where spread occupies » radio channel BW.

Now that the spread effective BW is significantly larger than the radio BW, the impact is markedly different. In this scenario, due to the time domain behavior of spread, the aggressor is now only in the victim radio channel for a proportionately small time. This can be calculated by simply dividing the radio BW by the spread BW, taking into account that the spread modulation completes a "round trip" for each modulation cycle (overlap occurs twice in a modulation cycle).

$$\text{Collision Probability (CP)} = \frac{2R_{BW}}{SSC_{BW}}$$

where $R_{BW} = $ Radio BW (MHz)

$SSC_{BW} = $ Spread BW (MHz) at target Frequency (F) given by $= F \times SSC\% \times 10^{-2}$

So for our example, the probability of collision (CP) is

$$CP = \frac{2 \times 0.2}{1940 \times 0.6 \times 10^{-2}} = \frac{0.4}{11.64} = 0.034.$$

This means that for every second, the aggressor frequency coincides (collides) with the victim radio channel for 34 msec. In this case, the addition of spread to the noise signal is clearly advantageous with the probability of a collision effectively reduced by a factor of 30.

So spread can either help or hinder you in pursuit of interference reduction. Understanding the fundamentals of spread can however help you both understand a potential impact and devise a possible mitigation strategy.

9.3 Radio Improvements

In Chapter 1, we discussed the fact that today radios in general are not designed to comprehend non-Gaussian noise sources. While not a major focus of this book, the subject of platform interference would not be complete without some discussion and direction for possible radio improvements in dealing with noise.

While older radio technologies have no way of distinguishing unwanted in-band signals from the intended signal, newer systems can incorporate several improvements that enhance their sensitivity. In digital radio systems, such as Wi-Fi, error-correction techniques can be used to improve performance. Spread-spectrum and frequency-hopping techniques can be also be used with digital signal processing techniques to improve resistance to interference. A highly directional receiver, such as a parabolic antenna or a diversity receiver, can be used to select one signal in space to the exclusion of others. What is missing from realizing these potential benefits is a fundamental knowledge of the aggressor and inclusion in the design parameters for the radio in question.

In order to highlight this potential, we'll consider the use of what is called an *autocorrelation function*. This is a well-known function in digital signal processing (DSP). Wikipedia defines this as follows:

> "Autocorrelation is a mathematical tool used frequently in signal processing for analyzing functions or series of values, such as time domain signals. Informally, it is a measure of how well a signal matches a time-shifted version of itself, as a function of the amount of time shift. More precisely, it is the cross-correlation of a signal with itself. Autocorrelation is useful for finding repeating patterns in a signal, such as determining the presence of a periodic signal which has been buried under noise, or identifying the missing fundamental frequency in a signal implied by its harmonic frequencies."

It sounds interesting, doesn't it? Typically this function is used as a method of identifying a wanted signal (normally periodic) that may have been "buried" in noise (normally white Gaussian). It can also be used to identify periodic noise signatures, and given that a large percentage of platform noise sources are periodic in nature, the same method can be applied to identifying noise. Given that these periodic noise signatures are in general static, it is then possible to use noise-cancellation techniques to subtract the noise from the total received RF signal.

We'll take this general direction and one specific case where some knowledge of the noise can dramatically improve radio performance in noisy environments.

9.3.1 Noise Estimation

OFDM is a multicarrier modulation method in which a wider radio spectrum is divided into narrower bands. This allows data to be essentially transmitted in parallel on these narrower bands. Signal-to-noise ratio (SNR) is defined as the ratio of desired signal power to the noise power and is the accepted standard measure of signal quality in communication systems. In adaptive radio systems, an estimate of SNR is required in order to be able to make best use of available resources. For example, in low or poor SNR environments, modification of the transmission parameters would be desired to compensate for poor channel conditions. In this way, such parameters as coding rate or modulation mode would be modified to satisfy system or application requirements such as a constant bit error rate

(BER), packet error rate (PER), or throughput. To achieve this in dynamic systems requires a real-time noise power estimator for continuous channel quality monitoring. With knowledge of SNR in digital multitone (DMT) systems, techniques such as adaptive bit loading can be used to improve performance by allocating more bits to subcarriers with higher SNRs. This same SNR knowledge also can be used for hand off algorithms, power control, and channel estimation through interpolation. This SNR can be established by using regular transmitted training sequences, pilot data, or data symbols, which is known as *blind estimation*. Alternatively, it can be estimated using actual data, but conventional techniques generally assume that all of the noise is white (Gaussian) and calculate a single SNR for all carriers. In order to benefit from bit loading, we need a per-carrier SNR. There has been much research published on this topic with numerous optimization techniques to establish a per-channel SNR. In recent years, minimal mean square error (MMSE) has emerged as the method of choice for establishing this per-carrier SNR due to its accuracy and ease of implementation. Using MMSE, white noise assumptions can be removed (a good thing because platform noise is not white) and therefore variation of noise power across OFDM subcarriers as well as across OFDM symbols can be incorporated.

To calculate this SNR, noise estimation is required. With per-carrier noise estimation, there is now an opportunity to improve the radio performance in the presence of our non-Gaussian platform noise. An example of such a system is shown in Figure 9.15.

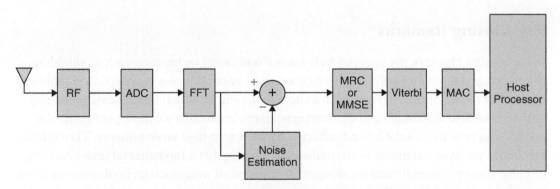

Figure 9.15: Block diagram of an interference mitigation system.

In this example, we simply use our noise estimation to subtract the noise from the signal prior to host processing. Since this function is done at each subcarrier, those without noise are unchanged while those with noise are essentially de-emphasized due to their higher noise content. This is also known as *tone puncturing*. Results from a basic simulation for such a system are shown in Figure 9.16.

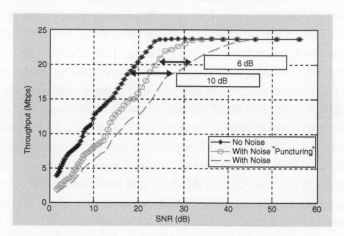

Figure 9.16: Selective tone puncturing simulation.

As we can see, without the tone puncturing, there is a 10 dB impact to the radio performance due to platform-generated noise. Adding the noise mitigation, approximately 6 dB of performance was regained.

9.4 Closing Remarks

In the closing chapters, we covered well-known traditional techniques such as shielding, absorbers, and layout as well as more recent source-based or noise suppression techniques. Given that most wireless devices such as notebooks and mobile Internet devices (MIDs) will need to deal with near-field electromagnetic environments, we have paid particular attention to how these solutions are affected by such near-field environments. Throughout this book, we have continued to strengthen the concept that a fundamental understanding of noise sources can enlighten the designer to know what mitigation technologies can work in what implementations. We have focused on understanding the sources and then the pursuit of mitigation. By following this approach, the engineer can avoid point solutions that work once and only once, forcing ground-up engineering in every design cycle.

We have also tried to approach the material in a somewhat different manner than is usual by providing quite a bit of visual content. Most textbooks provide all of the equations and then discuss the equations. We have tried to develop for the reader an approach that allows for building an intuitive understanding of the methods, the models, and the measurements. We want the reader to come away believing that he or she can use these methods to explore how varying parameters can change the results. We also want the reader to develop a grasp of how each individual parameter will have an impact on the problem. As we have repeatedly stated, there are many mitigation techniques. Some provide many decibels of improvement, and some only a couple of decibels. The reader should know when each technique is applicable and remember that a set of small contributions can add to the same impact as one large contribution, making your mitigation techniques easier and less costly to implement.

References

[1] C. Paul, *Introduction to Electromagnetic Compatibility* (2nd ed.), Wiley, 2006.

[2] H.W. Ott, *Noise Reduction Techniques in Electronic Systems* (2nd ed.), Wiley, 1988.

[3] M. Mardiguian, *Controlling Radiated Emissions by Design,* Springer, 2001.

[4] D.R.J. White, *Electromagnetic Interference and Compatibility,* vol. 3, *EMI Control Methods and Techniques*, Don White Consultants, 1973.

[5] A.F. Molisch, *Wireless Communications,* Wiley, 2005.

[6] TDK Application Note: *EMC Components Radio Wave Absorbers: Application Examples of Electromagnetic Absorbers* (Ref 001-03/20070302/e9e_bdj_003). *EMI/RFI Shielding Frequently Asked Questions,* Thermospray Company.

[7] W.D. Kimmel, and D.D. Gerke, *EMC Design for Compliance: Conductive Coatings* (Conformity Whitepaper).

[8] A. Sundsmo, *Microwave Absorbers: Reducing Cavity Resonances,* Laird Technologies (Compliance Engineering Whitepaper).

[9] J. Armstrong and H.A. Suraweera, *Impulse Noise Mitigation for OFDM Using Decision Directed Noise Estimation.* ISSSTA2004, Sydney, Australia (30 Aug to 2 Sept 2004).

[10] T. Yucek and H. Arslan, *MMSE Noise Power and SNR Estimation for OFDM Systems.* (Whitepaper, University of South Florida).

[11] A.N. Barreto and S. Furrer, *Adaptive Bit Loading for Wireless OFDM Systems.* IBM Zurich Research Laboratory, Ruschlikon, Switzerland.

[12] K.M. Nasr, F. Costen, and S.K. Barton, *Performance of Different Interpolation Strategies for OFDM/MMSE Smart Antenna System in an Indoor WLAN.* The University of Manchester, UK. Vehicular Technology Conference, 2005.

[13] Y. Kakishima, H. Le, S.H. Ting, K. Sakaguchi, and K. Araki, *Experimental Analysis of MIMO-OFDM Eigenmode Transmission with MMSE Interference Canceller.* Tokyo Institute of Technology. The 17th Annual IEEE International Symposium on Personal, Indoor and Mobile Radio Communication (PIMRC 2006).

Appendix to Chapter 2

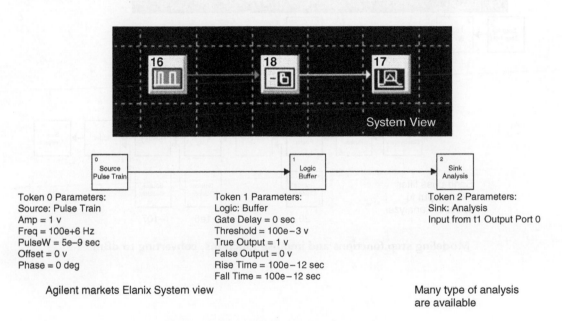

Token 0 Parameters:
Source: Pulse Train
Amp = 1 v
Freq = 100e+6 Hz
PulseW = 5e–9 sec
Offset = 0 v
Phase = 0 deg

Token 1 Parameters:
Logic: Buffer
Gate Delay = 0 sec
Threshold = 100e–3 v
True Output = 1 v
False Output = 0 v
Rise Time = 100e–12 sec
Fall Time = 100e–12 sec

Token 2 Parameters:
Sink: Analysis
Input from t1 Output Port 0

Agilent markets Elanix System view

Many type of analysis
are available

System view model for analysis of signals.

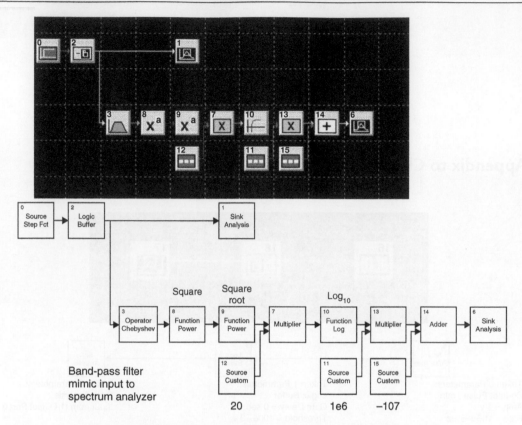

Band-pass filter
mimic input to
spectrum analyzer

Modeling step functions and impulse functions, converting to dBm.

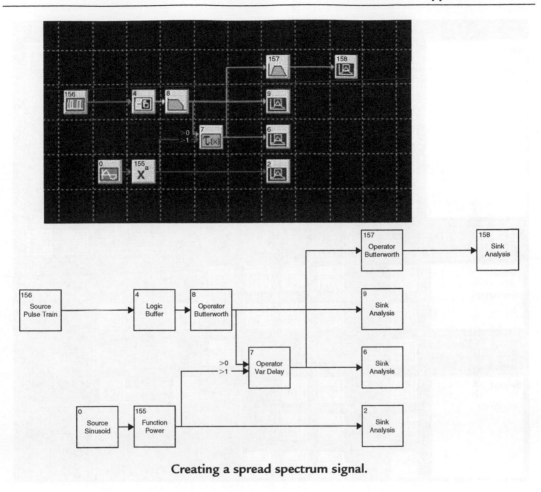

Creating a spread spectrum signal.

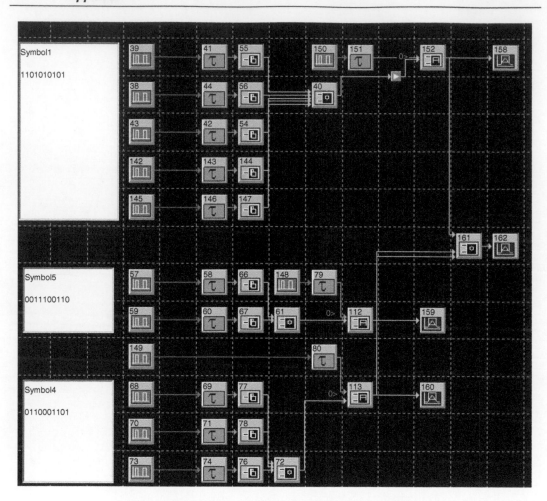

Time and frequency behavior of complex symbols.

$\text{Clear}[t1, t2, t3, t4, t5, t6, t7, t8, t9, t10, t11, t12, t, a1, a2, a3, n, T, f2, A0, Df2];$

$f2[t1\_, t2\_, t3\_, t4\_, t5\_, t6\_, t7\_, t8\_, t9\_, t10\_, t11\_, t12\_, T\_, a1\_, a2\_, a3\_]$

$$= \int_{t1}^{t2} a1\left(1 - e^{-(t-t1)/\left(\frac{t2-t1}{5}\right)}\right) e^{\left(-\frac{2\pi i n t}{T}\right)} dt + \int_{t2}^{t3} a1 e^{\left(-\frac{2\pi i n t}{T}\right)} dt + \int_{t3}^{t4} a1\left(e^{-(t-t3)/\left(\frac{t4-t3}{5}\right)}\right) e^{\left(-\frac{2\pi i n t}{T}\right)} dt$$

$$+ \int_{t5}^{t6} a2\left(1 - e^{-(t-t5)/\left(\frac{t6-t5}{5}\right)}\right) e^{\left(-\frac{2\pi i n t}{T}\right)} dt + \int_{t6}^{t7} a2 e^{\left(-\frac{2\pi i n t}{T}\right)} dt + \int_{t7}^{t8} a2\left(e^{-(t-t7)/\left(\frac{t8-t7}{5}\right)}\right) e^{\left(-\frac{2\pi i n t}{T}\right)} dt$$

$$+ \int_{t9}^{t10} a3\left(1 - e^{-(t-t9)/\left(\frac{t10-t9}{5}\right)}\right) e^{\left(-\frac{2\pi i n t}{T}\right)} dt + \int_{t10}^{t11} a3 e^{\left(-\frac{2\pi i n t}{T}\right)} dt + \int_{t11}^{t12} a3\left(e^{-(t-t11)/\left(\frac{t12-t11}{5}\right)}\right) e^{\left(-\frac{2\pi i n t}{T}\right)} dt;$$

$$A0 = \frac{1}{T}\left(\int_{t1}^{t2} a1\left(1 - e^{-(t-t1)/\left(\frac{t2-t1}{5}\right)}\right) e^{(0)} dt + \int_{t2}^{t3} a1 e^{(0)} dt + \int_{t3}^{t4} a1\left(e^{-(t-t3)/\left(\frac{t4-t3}{5}\right)}\right) e^{(0)} dt\right.$$

$$+ \int_{t5}^{t6} a2\left(1 - e^{-(t-t5)/\left(\frac{t6-t5}{5}\right)}\right) e^{(-0)} dt + \int_{t6}^{t7} a2 e^{(-0)} dt + \int_{t7}^{t8} a2\left(e^{-(t-t7)/\left(\frac{t8-t7}{5}\right)}\right) e^{(-0)} dt$$

$$\left. + \int_{t9}^{t10} a3\left(1 - e^{-(t-t9)/\left(\frac{t10-t9}{5}\right)}\right) e^{(-0)} dt + \int_{t10}^{t11} a3 e^{(-0)} dt + \int_{t11}^{t12} a3\left(e^{-(t-t11)/\left(\frac{t12-t11}{5}\right)}\right) e^{(-0)} dt\right);$$

$Df2[t1\_, t2\_, t3\_, t4\_, t5\_, t6\_, t7\_, t8\_, t9\_, t10\_, t11\_, t12\_, T\_, a1\_, a2\_, a3\_]$

$$= \int_{t1}^{t2} D\left[a1\left(1 - e^{-(t-t1)/\left(\frac{t2-t1}{5}\right)}\right), t\right] e^{\left(\frac{2\pi i n t}{T}\right)} dt + \int_{t3}^{t4} D\left[a1\left(e^{-(t-t3)/\left(\frac{t4-t3}{5}\right)}\right), t\right] e^{\left(\frac{2\pi i n t}{T}\right)} dt$$

$$+ \int_{t5}^{t6} D\left[a2\left(1 - e^{-(t-t5)/\left(\frac{t6-t5}{5}\right)}\right), t\right] e^{\left(-\frac{2\pi i n t}{T}\right)} dt + \int_{t7}^{t8} D\left[a2\left(e^{-(t-t7)/\left(\frac{t8-t7}{5}\right)}\right), t\right] e^{\left(-\frac{2\pi i n t}{T}\right)} dt$$

$$+ \int_{t9}^{t10} D\left[a3\left(1 - e^{-(t-t9)/\left(\frac{t10-t9}{5}\right)}\right), t\right] e^{\left(-\frac{2\pi i n t}{T}\right)} dt + \int_{t11}^{t12} D\left[a3\left(e^{-(t-t11)/\left(\frac{t12-t11}{5}\right)}\right), t\right] e^{\left(-\frac{2\pi i n t}{T}\right)} dt;$$

Mathematica code for Fourier components of a symbol.

```
Clear[vsig, v, v2, coeff, data2, T, a1, a2, a3, t1, t2, t3, t4, t5, t6, t7, t8, t9, t10, t11, t12, p, n, t]
```

frequency $= 100\ 10^6$; $p = 50$;

$$T = \frac{1}{\text{frequency}};$$

$t1 = 0\ 10^{-9}$;
$t2 = (0.1\ 10^{-9} + \text{Random}[\text{Real}, \{-40\ 10^{-12}, 40\ 10^{-12}\}])$;
$t3 = (5.0\ 10^{-9} + \text{Random}[\text{Real}, \{-50\ 10^{-12}, 50\ 10^{-12}\}])$;
$t4 = (5.1\ 10^{-9} + \text{Random}[\text{Real}, \{-40\ 10^{-12}, 40\ 10^{-12}\}])$;
$t5 = 0\ 10^{-9}$;

$t6 = 1\ 10^{-9}$;

$t7 = 5\ 10^{-9}$;
$t8 = 6\ 10^{-9}$;
$t9 = 8\ 10^{-9}$;
$t10 = 8.1\ 10^{-9}$;
$t11 = 9\ 10^{-9}$;
$t12 = 9.1\ 10^{-9}$;
$a1 = 0$; $a2 = 1$; $a3 = 0$;

$$\text{coeff}[t\_, n\_] = \text{Exp}\left[\frac{I2\pi nt}{T}\right];$$

$$v[t\_] = A0 + \left(\sum_{n=1}^{p} \frac{1}{T}\left(2\,\text{coeff}[t, n]\,f2[t1, t2, t3, t4, t5, t6, t7, t8, t9, t10, t11, t12, T, a1, a2, a3]\right)\right);$$

$$v2[t\_] = \left(\frac{1}{T}\left(2\,\text{coeff}[t, n]\,f2[t1, t2, t3, t4, t5, t6, t7, t8, t9, t10, t11, t12, T, a1, a2, a3]\right)\right);$$

$$\text{vsig}[T\_, n\_] = \frac{1}{T}\left(f2[t1, t2, t3, t4, t5, t6, t7, t8, t9, t10, t11, t12, T, a1, a2, a3]\right);$$

$$\text{Dv}[t\_] = \left(\sum_{n=1}^{p} \frac{1}{T}\left(2\,\text{coeff}[t, n]\,Df2[t1, t2, t3, t4, t5, t6, t7, t8, t9, t10, t11, t12, T, a1, a2, a3]\right)\right);$$

(*

$$\text{data3} = \text{Table}\left[\frac{1}{T}\left(\sqrt{2}\,f2[t1, t2, t3, t4, t5, t6, t7, t8, t9, t10, t11, t12, T, a1, a2, a3]\right), \{n, 1, p, 1\}\right];$$

$$\text{data3D} = \text{Table}\left[\frac{1}{T}\left(\sqrt{2}\,Df2[t1, t2, t3, t4, t5, t6, t7, t8, t9, t10, t11, t12, T, a1, a2, a3]\,100\ 10^{-12}\right), \{n, 1, p, 1\}\right];$$

Print [" time domain reconstruction of the Real part of the signal "]
Plot [Re[$v[t]$], {t, $-1\,10^{-9}$, $10\,10^{-9}$}, PlotRange → All]

Plot [Re[$Dv[t]$] $100\,10^{-12}$, {t, $-1\,10^{-9}$, $9.99\,10^{-9}$}, PlotRange → All]

Print[" time domain reconstruction of the imaginary part of the signal "]
Plot[Im[$Dv[t]$] $100\,10^{-12}$, {t, $-1\,10^{-9}$, $9.99\,10^{-9}$}, PlotRange → All]
(*
 Print [" frequency domain reconstruction of the $|$Im$|$ signal "]

plotg21 = Plot$\left[20\,\text{Log}\left[10, \text{Abs}\left[\text{Im}\left[\frac{\sqrt{2}f2[t1,t2,t3,t4,t5,t6,t7,t8,t9,t10,t11,t12,T,a1,a2,a3]}{T} \right] \right] 1\,10^6 \right] - 108, \right.$

$\quad$ {$n,5,7$}, PlotRange → {$-80,10$}, PlotStyle → {RGBColor[0, .5, 1],Thickness[.004]}$\Big]$

Print [" frequency domain reconstruction of the $|$Re$|$ signal "]

ploth21 = Plot$\left[20\,\text{Log}\left[10, \text{Abs}\left[\text{Re}\left[\frac{\sqrt{2}f2[t1,t2,t3,t4,t5,t6,t7,t8,t9,t10,t11,t12,T,a1,a2,a3]}{T} \right] \right] 1\,10^6 \right] - 108, \right.$

$\quad$ {$n,5,7$}, PlotRange→ {$-80,10$}, PlotStyle → {RGBColor[.5, .5, 0], Thickness[.004]}$\Big]$

Print [" frequency domain reconstruction of the $|$Abs$|$ signal "]

plotf = Plot$\left[20\,\text{Log}\left[10, \text{Abs}\left[\frac{\sqrt{2}f2[t1,t2,t3,t4,t5,t6,t7,t8,t9,t10,t11,t12,T,a1,a2,a3]}{T} \right] 1\,10^6 \right] - 108, \right.$

$\quad$ {$n,1,p$}, PlotRange → {$-80,10$}$\Big]$

Appendix to Chapters 4–7

As in the previous appendix, we give here a set of analytical models and system simulation models to enable the reader to pursue research and investigation independently. No attempt at elegance in the code has been made. Since memory and computer speed are so high these days, there isn't any point in it anymore. So here they are. They work, and they have provided much of the material in this chapter. The reader will have to transcribe them independently. Have fun!

The following lines of Mathematical code describe the elementary point dipole equations. The form of the equations is directly from Clayton Paul's *Introduction to Electromagnetics*, which, by the way, is one of the best books of its kind and is highly recommended.

(* electric dipole *)

$$\text{eFieldr}[r\_,\text{theta}\_] = 2\frac{\text{cur}\,dl}{4\pi}\,\text{eta0}\,\text{beta0}^2\cos[\text{theta}]\left(\frac{1}{\text{beta0}^2\,r^2} - i\frac{1}{\text{beta0}^3\,r^3}\right)e^{-i\,\text{beta0}r};$$

$$\text{eField}\theta[r\_,\text{theta}\_] = \frac{\text{cur}\,dl}{4\pi}\,\text{eta0}\,\text{beta0}^2\sin[\text{theta}]\left(i\frac{1}{\text{beta0}\,r} + \frac{1}{\text{beta0}^2\,r^2} - i\frac{1}{\text{beta0}^3\,r^3}\right)e^{-i\,\text{beta0}r};$$

$$\text{hField}\varphi[r\_,\text{theta}\_] = \frac{\text{cur}\,dl}{4\pi}\,\text{beta0}^2\sin[\text{theta}]\left(i\frac{1}{\text{beta0}\,r} + \frac{1}{\text{beta0}^2\,r^2}\right)e^{-i\,\text{beta0}r};$$

(* magnetic dipole *)

$$\text{hFieldr}[r\_,\text{theta}\_] = 2\frac{i\,\text{omegamu0megmoment}}{4\pi\text{eta0}}\,\text{beta0}^2\cos[\text{theta}]\left(\frac{1}{\text{beta0}^2\,r^2} - i\frac{1}{\text{beta0}^3\,r^3}\right)e^{-i\,\text{beta0}r};$$

$$\text{hField}\theta[r\_,\text{theta}\_] = \frac{i\,\text{omegamu0megmoment}}{4\pi\text{eta0}}\,\text{beta0}^2\sin[\text{theta}]\left(i\frac{1}{\text{beta0}\,r} + \frac{1}{\text{beta0}^2\,r^2} - i\frac{1}{\text{beta0}^3\,r^3}\right)e^{-i\,\text{beta0}r};$$

$$\text{eField}\varphi[r\_,\text{theta}\_] = \frac{-i\,\text{omegamu0megmoment}}{4\pi}\,\text{beta0}^2\sin[\text{theta}]\left(i\frac{1}{\text{beta0}\,r} + \frac{1}{\text{beta0}^2\,r^2}\right)e^{-i\,\text{beta0}r};$$

$$\text{wavez}[r\_,\text{theta}\_] = \sqrt{\left(\frac{\text{eFieldr}\,[r,\text{theta}]}{\text{hField}\varphi\,[r,\text{theta}]}\right)^2 + \left(\frac{\text{eField}\theta\,[r,\text{theta}]}{\text{hField}\varphi\,[r,\text{theta}]}\right)^2}\qquad\text{For the electric dipole}$$

Mathematica code for point source equations.

$\text{ind} = 5\,10^{-9}; \text{cap} = 10\,10^{-12}; (* \text{ind} = 5, \text{cap} = 20 *)$

$$\text{curE} = \frac{.005}{\sqrt{\frac{1 + l2\pi\,\text{frequency ind}}{5}}}; dl = .0001; \left(* \text{ for B probe use } \sqrt{\frac{5 + 2\pi\,\text{frequency ind}}{1 + 2\pi\,\text{frequency cap}}} \text{ with ind} = 0 *\right)$$

$$\text{curM} = \frac{.07}{\sqrt{\frac{5}{l2\pi\,\text{frequency cap}}}};$$

$\text{frequency} = 3000\,10^6;$

$\text{radius} = .0005;$

$$\text{lambda} = N\left[\frac{299.8\,10^6}{\text{frequency}}\right];$$

$\text{megmoment} = N[\text{curM}\pi\,\text{radius}^2];$

$$\text{beta0} = N\left[\frac{2\pi}{\text{lambda}}\right];$$

$\text{omega} = N[2\pi\text{frequency}];$

$\text{eta0} = 377; e0 = \frac{8.854}{10^{12}}; \text{er} = 1; \text{mur} = 1;$

$$\text{mu0} = N\left[\frac{4\pi}{10^7}\right];$$

$$\text{const} = N\left[\frac{\text{mu0murmegmoment}\,\pi}{4}\right];$$

The following lines of code are what we used to produce the results on skew and the statistical distributions of rise time, fall time, and pulse width jitter.

$curr = \dfrac{.00007}{50}$; (* Gen 1 amplitude at 2.45 GHz, single lane at xmtr, rbw = 10 Khz *)

$freq = 2.45\,10^9$; $er = 8$;

$lambda = \dfrac{\frac{300\,10^6}{\sqrt{er}}}{freq}$;

$beta = \dfrac{2\pi}{lambda}$;

$skewM = 50$;

$d1 = .001$; (* 1/20 wavelength at 3 GHz *)

$Etheta\,[r\_] = \dfrac{curr\,d1}{4\pi}\,377\,beta^2 Sin\left[\left(\dfrac{i}{beta\,r} + \dfrac{1}{beta^2 r^2} - \dfrac{i}{beta^3 r^3}\right)\right] e^{-i\,beta\,r}$;

$Er\,[r\_] = \dfrac{2\,curr\,d1}{4\pi}\,377\,beta^2 Cos\left[\left(\dfrac{1}{beta^2 r^2} - \dfrac{i}{beta^3 r^3}\right)\right] e^{-i\,beta\,r}$;

$Ecomb[r\_] = \sqrt{(Etheta[r]^2 + Er[r]^2)}$;

$elp[n\_, phase\_, skew\_, xi\_] = Abs[Etheta[rp]]\,Sin[2\pi\,freq + phase]$;

$eln[n\_, phase\_, skew\_, xi\_] = Abs[Etheta[rn]]\,Sin\left[2\pi\,freq + phase + \pi + \frac{skew\pi}{400} + \frac{xi}{400}\right]$;

(* gen 1 = 400 , gen 2 = 200 *)

```
Etotal = 0; Elist = {};
For[p = 1, p < 1000,
    Etotal = 0;
    For[q = 1, q < 9,
    clear [phase, skew, deltaskew, xi];
```

rp = .01 + Random [Real, {0, .002}]; rn = $\sqrt{rp^2 + d1^2}$;
(* vary distance with number of pairs *)

deltaskew = Random $\left[\text{Real, } \left\{ -\dfrac{\text{skew}M}{.05}, \dfrac{\text{skew}M}{.05} \right\} \right]$;

phase = Random [Real, {0, 2π}];

skew = skewM + deltaskew;   (* channel skew *)

For[n = 1, n < 2, (* 2 is a single channel, single point, 4 is single channel 3 points *)

xi = 10 + Random [Real, {−5,5}];   (* inter pair skew *)

Etotal = Etotal + elp[n, phase, skew, xi] + eln[n, phase, skew, xi];

(* Print [skew," ", deltaskew, " ",xi] *),

n++]; Elist = Append[Elist, Etotal], q++], p++];

Histogram [20 Log [10, Abs [Elist] 1000000] −108, HistogramScale → 1]

The next piece is what was used to produce the analysis of compact complex sets of radiative point sources and the creation of radiation surfaces. This one especially is why we mentioned that we made no attempt at elegance, but keep at it until it actually produced correct results.

```
stepsize = 100 10⁻⁶; h = .0001;
xs1 = 5; ys1 = 25; xs2 = 15; ys2 = 25; xs3 = 25; ys3 = 25;
xs4 = 5; ys4 = 15; xs5 = 15; ys5 = 15; xs6 = 25; ys6 = 15;
xs7 = 5; ys7 = 5;  xs8 = 15; ys8 = 5;  xs9 = 25; ys9 = 5; xs10 = 16; ys10 = 16;
p = 30;
EdataHi = Array [data,  {p, p}];
EdataEi = Array [data,  {p, p}];
EdataHiph = Array [data,  {p, p}];
EdataEiph = Array [data,  {p, p}];
For [i = 1, i < p + 1];
  For [j =1, j < p + 1];
    sourcex1 = stepsize xs1; sourcex2 = stepsize xs2; sourcex3 = stepsize xs3;
    sourcey1 = stepsize ys1; sourcey2 = stepsize ys2; sourcey3 = stepsize ys3;
    sourcex4 = stepsize xs4; sourcex5 = stepsize xs5;  sourcex6 = stepsize xs6;
    sourcey4 = stepsize ys4; sourcey5 = stepsize ys5; sourcey6 = stepsize ys6;
    sourcex7 = stepsize xs7; sourcex8 = stepsize xs8; sourcex9 = stepsize xs9;
    sourcey7 = stepsize ys7; sourcey8 = stepsize ys8; sourcey9 = stepsize ys9;

    dx1 = sourcex1 − stepsize i; dy1 = sourcey1 − stepsize j;

    dx2 = sourcex2 − stepsize i; dx3 = sourcex3 − stepsize i;
    dy2 = sourcey2 − stepsize j; dy3 = sourcey3 − stepsize j;
    dx4 = sourcex4 − stepsize i; dx5 = sourcex5 − stepsize i; dx6 = sourcex6 − stepsize i;
    dy4 = sourcey4 − stepsize j; dy5 = sourcey5 − stepsize j; dy6 = sourcey6 − stepsize j;
    dx7 = sourcex7 − stepsize i; dx8 = sourcex8 − stepsize i; dx9 = sourcex9 − stepsize i;
    dy7 = sourcey7 − stepsize j; dy8 = sourcey8 − stepsize j; dy9 = sourcey9 − stepsize j;
```

$r11 = \sqrt{dy1^2 + dx1^2}$; $r21 = \sqrt{h^2 + r11^2}$; If[stepsize i > sourcex1, $r21 = r21\,(-1)$, $r21 = r21$];

$r12 = \sqrt{dy2^2 + dx2^2}$; $r22 = \sqrt{h^2 + r12^2}$; If[stepsize i > sourcex2, $r22 = r22\,(-1)$, $r22 = r22$];

$r13 = \sqrt{dy3^2 + dx3^2}$; $r23 = \sqrt{h^2 + r13^2}$; If[stepsize i > sourcex3, $r23 = r23\,(-1)$, $r23 = r23$];

$r112 = \sqrt{dy4^2 + dx4^2}$; $r212 = \sqrt{h^2 + r112^2}$; If[stepsize i > sourcex4, $r212 = r212\,(-1)$, $r212 = r212$];

$r122 = \sqrt{dy5^2 + dx5^2}$; $r222 = \sqrt{h^2 + r122^2}$; If[i > xs5, $r222 = r222\,(-1)$, $r222 = r222$];

$r132 = \sqrt{dy6^2 + dx6^2}$; $r232 = \sqrt{h^2 + r132^2}$; If[stepsize i > sourcex6, $r232 = r232\,(-1)$, $r232 = r232$];

$r113 = \sqrt{dy7^2 + dx7^2}$; $r213 = \sqrt{h^2 + r113^2}$; If[stepsize i > sourcex7, $r213 = r213\,(-1)$, $r213 = r213$];

$r123 = \sqrt{dy8^2 + dx8^2}$; $r223 = \sqrt{h^2 + r123^2}$; If[stepsize i > sourcex8, $r223 = r223\,(-1)$, $r223 = r223$];

$r133 = \sqrt{dy9^2 + dx9^2}$; $r233 = \sqrt{h^2 + r133^2}$; If[stepsize i > sourcex9, $r233 = r233\,(-1)$, $r233 = r233$];

$r143 = \sqrt{dy10^2 + dx10^2}$; $r243 = \sqrt{h^2 + r143^2}$; If[stepsize i > sourcex10, $r243 = r243\,(-1)$, $r243 = r243$];

$ph = 2$;

randphase1 = Random $\left[\text{Real}, \left\{-\dfrac{\pi}{ph}, \dfrac{\pi}{ph}\right\}\right]$; randphase4 = Random $\left[\text{Real}, \left\{-\dfrac{\pi}{ph}, \dfrac{\pi}{ph}\right\}\right]$;

randphase2 = Random $\left[\text{Real}, \left\{-\dfrac{\pi}{ph}, \dfrac{\pi}{ph}\right\}\right]$; randphase5 = Random $\left[\text{Real}, \left\{-\dfrac{\pi}{ph}, \dfrac{\pi}{ph}\right\}\right]$;

randphase3 = Random $\left[\text{Real}, \left\{-\dfrac{\pi}{ph}, \dfrac{\pi}{ph}\right\}\right]$; randphase6 = Random $\left[\text{Real}, \left\{-\dfrac{\pi}{ph}, \dfrac{\pi}{ph}\right\}\right]$;

randphase7 = Random $\left[\text{Real}, \left\{-\dfrac{\pi}{ph}, \dfrac{\pi}{ph}\right\}\right]$;

randphase8 = Random $\left[\text{Real}, \left\{-\dfrac{\pi}{ph}, \dfrac{\pi}{ph}\right\}\right]$;

randphase9 = Random $\left[\text{Real}, \left\{-\dfrac{\pi}{ph}, \dfrac{\pi}{ph}\right\}\right]$;

randphase10 = Random $\left[\text{Real}, \left\{-\dfrac{\pi}{ph}, \dfrac{\pi}{ph}\right\}\right]$;

$$\text{theta}1 = \frac{(i \text{ stepsize} - (\text{sourcex1}))}{r21} + \frac{\pi}{2};$$

$$\text{theta}2 = \frac{(i \text{ stepsize} - (\text{sourcex2}))}{r22} + \frac{\pi}{2};$$

$$\text{theta}3 = \frac{(i \text{ stepsize} - (\text{sourcex3}))}{r23} + \frac{\pi}{2};$$

$$\text{theta}4 = \frac{(i \text{ stepsize} - (\text{sourcex4}))}{r212} + \frac{\pi}{2};$$

$$\text{theta}5 = \frac{(i \text{ stepsize} - (\text{sourcex5}))}{r222} + \frac{\pi}{2};$$

$$\text{theta}6 = \frac{(i \text{ stepsize} - (\text{sourcex6}))}{r232} + \frac{\pi}{2};$$

$$\text{theta}7 = \frac{(i \text{ stepsize} - (\text{sourcex7}))}{r213} + \frac{\pi}{2};$$

$$\text{theta}8 = \frac{(i \text{ stepsize} - (\text{sourcex8}))}{r223} + \frac{\pi}{2};$$

$$\text{theta}9 = \frac{(i \text{ stepsize} - (\text{sourcex9}))}{r233} + \frac{\pi}{2};$$

$$\text{theta}10 = \frac{(i \text{ stepsize} - (\text{sourcex10}))}{r243} + \frac{\pi}{2};$$

noiseM = 0; (∗ 50 works well in close ∗)

EdataHi[[i, j]] = noiseM RandomNormal[0.5, 0.25]

$+\sqrt{(}$(hFieldφ[r21, theta1]^1+hFieldφ[r22, theta2]^1+hFieldφ[r23, theta3]^1+ hFieldφ[r212, theta4]^1

+ hFieldφ[r222, theta5]^1+hFieldφ[r232, theta6]^1+hFieldφ[r213, theta7]^1+hFieldφ[r223, theta8]^1

+ hFieldφ[r233, theta9]^1)^2);

EdataEi[[i, j]] = noiseM RandomNormal[0.5, 0.25]

$+\sqrt{(}$(eFieldφ[r21, theta1]^1+eFieldφ[r22, theta2]^1+eFieldφ[r23, theta3]^1+eFieldφ[r212, theta4]^1

+ eFieldφ[r222, theta5]^1+eFieldφ[r232, theta6]^1+eFieldφ[r213, theta7]^1+eFieldφ[r223, theta8]^1

+ eFieldφ[r233, theta9]^1)^2);

EdataHiph[[i, j]] = noiseM RandomNormal[0.5, 0.25]

$+\sqrt{(}$(hFieldφ[r21, theta1+randphase1]^1+hFieldφ[r22, theta2 + randphase2]^1+hFieldφ[r23, theta3 + randphase3]^1

+ hFieldφ[r212, theta4 + randphase4]^1+hFieldφ[r222, theta5 + randphase5]^1

+ hFieldφ[r232, theta6 + randphase6]^1+hFieldφ[r213, theta7 + randphase7]^1

+ hFieldφ[r223, theta8 + randphase8]^1+hFieldφ[r233, theta9 + randphase9]^1)^2);

EdataEiph[[i, j]] = noiseM RandomNormal[0.5, 0.25]

$+\sqrt{(}$(eFieldφ[r21, theta1+randphase1]^1+eFieldφ[r22, theta2+randphase2]^1+eFieldφ[r23, theta3 + randphase3]^1

+ eFieldφ[r212, theta4 + randphase4]^1+eFieldφ[r222, theta5+randphase5]^1

+ eFieldφ[r232, theta6+randphase6]^1+eFieldφ[r213, theta7+randphase7]^1

+ eFieldφ[r223, theta8 + randphase8]^1+eFieldφ[r233, theta9+randphase9]^1)^2); j++]; i++];

$EdataH = \dfrac{Re[EdataHi]}{1};$

$EdataE = \dfrac{Re[EdataEi]}{1};$

$EdataHph = \dfrac{EdataHiph}{Max[Abs[EdataHiph]]};$

$EdataEph = \dfrac{EdataEiph}{Max[Abs[EdataEiph]]};$

xgrid = Table[i, {i, 1, p}];
ygrid = Table[j, {j, 1, p}];

n = 1;
matrixSIZE = p; (* matrix must be square and even *)
mask = Table [If[$(i - (\text{matrixSIZE}/2 + 1))\hat{} 2 + (j - (\text{matrixSIZE}/2 + 1))\hat{} 2 < 250$,

$$\frac{1}{1 + \left(\frac{\sqrt{(i - (\text{matrixSIZE}/2+1))^2 + (j - (\text{matrixSIZE}/2+1))^2}}{10}\right)^8}, 0],$$

{i, matrixSIZE}, {j, matrixSIZE}];

(* low frequency filter $\dfrac{15}{1\,10^{-9} + \sqrt{(i - (\text{matrixSize}/2 + 1))^2 + (j - (\text{matrixSIZE}/2 + 1))^2}}$ high frequency filter

$\dfrac{\sqrt{(i - (\text{matrixSize}/2 + 1))^2 + (j - (\text{matrixSIZE}/2 + 1))^2}}{15}$ *)

recons1E = Chop [IDFT[

        Map[RotateRight[#, matrixSIZE/2] &, RotateRight[mask, matrixSIZE/2]] DFT[EdataE]]];

recons1H = Chop [IDFT[

        Map[RotateRight[#, matrixSIZE/2] &, RotateRight[mask, matrixSIZE/2]] DFT[EdataH]]];

recons1Eph = Chop [IDFT[

        Map[RotateRight[#, matrixSIZE/2] &, RotateRight[mask, matrixSIZE/2]] DFT[EdataEph]]];

recons1Hph = Chop [IDFT[

        Map[RotateRight[#, matrixSIZE/2] &, RotateRight[mask, matrixSIZE/2]] DFT[EdataHph]]];

We finish with the Mathematica model for power plane resonances. The first set of lines describes the determination of the frequency of resonance.

```
Clear[m, n, len, w, er, volt, resFREQ, Kx, Ky, h]

(* modes in a power plate assembly h = plane separation *)

len = .070; w = .055; h = .00040; m = 2; n = 3; ka = 2;

(* ka = 2 for m = n; ka = 4 for m NEQ n *)
```

$$kx[m\_] = \frac{m\pi}{\text{len}};$$

$$ky[n\_] = \frac{n\pi}{w};$$

$$c0 = 300 \ 10^6 \ ; \ er = 4.1;$$

```
volt[x_, y_] = Cos[ kx[m]x] Cos[ky[n]y];
```

$$\text{resFreq}[m\_, n\_] = \frac{c0}{2\pi\sqrt{er}}\sqrt{\left(\frac{m\pi}{\text{len}}\right)^2 + \left(\frac{n\pi}{w}\right)^2};$$

```
Print["resonant frequency = ", resFreq[m, n]]

Plot3D[Abs[volt[x, y]], {x, 0, len}, {y, 0, w}, PlotPoints → 50,

BoxRatios → {60, 75, 10}]
```

The next lines describe the actual excitation distribution.

Clear[x, y, kx, ky, volt, v, volt2, omega]

$x0 = .03; y0 = .005;$

$kx[m\_] = \dfrac{m\pi}{len};$

$ky[n\_] = \dfrac{n\pi}{w};$

$c0 = 300 \, 10^6;$ er $= 4.5;$ losstangent $= .025;$ sigma $= 5.65 \, 10^7;$ mu0 $= 4\pi \, 10^{-7};$

omega $= 2\pi$ resFreq[m, n];

$$quality[omega\_] = \dfrac{1}{losstangent + \dfrac{1}{h}\sqrt{\frac{2}{omega \, mu0 \, sigma}}};$$

volt[x_, y_] = Cos[kx[m]x]Cos[ky[n]y];

$$v[m\_, n\_] = \dfrac{120}{\sqrt{er}}\dfrac{h}{w}\dfrac{ka \, Cos[kx[m]x0]Cos[ky[n]y0]}{\sqrt{m^2 + n^2 \left(\frac{len}{w}\right)^2}} \, quality \, [omega];$$

volt2[x_, y_] = v[m, n] volt [x, y];

Plot3D[Abs[volt2[x,y]], {x, 0, len}, {y, 0, w}, PlotPoints → 50,

BoxRatios → {55, 74, 10}]

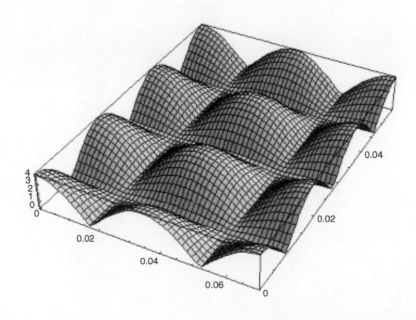

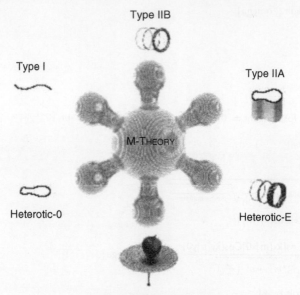

Type IIB

Type I

M-Theory

Type IIA

Heterotic-0

Heterotic-E

11 dimensional supergravity

11-dimensional supergravity.

Index

A

Absorbers, 276–281
  applications of, 277
  characteristics, 278
  disadvantage of, 278
  in LCD panel, 279
  microwave, 278
Absorption loss, 256, 276
  of copper and aluminum,
    259
Access point (AP), 9
Active mitigation
  frequency content, 297–301
  frequency planning, 291–297
  radio improvements, 305
ADC; *see* Analog-to-digital
  converter
Additive noise, 17–18
Additive white Gaussian noise
  (AWGN), 14
Ampere's law, 153
Analog-to-digital converter
  (ADC), 8
  with and without platform
    noise, 9
Analytical surface operators, 149
Antenna placement
  LCD panel, 282
  noise reduction from, 284
  WLAN, 283, 286
  WWAN, 285
AP; *see* Access point

Asymmetric clock signal, 43, 113
  even harmonics, 45, 78
  and symmetric clock, phase
    shift of, 84–85
Autocorrelation function, 306
AWGN; *see* Additive white
  Gaussian noise

B

Band-pass filter, 73
BER; *see* Bit error rate
Bit error rate (BER), 20, 306
10 Bit symbol, 66–67, 88, 101
  spectra of, 78–79
Broadband noise, 13

C

Cables, 213
Capacitor plates, 154–156
CDMA; *see* Code division
  multiple access
Centrino®, 2
Channel radiation model, 203–204
Channel skew, 205–208, 210
CK505, 133
Clayton Paul's model, 181, 206
Clock signal
  bit structure for, 77, 80
  differentiated, spectrum of, 96
  and Fourier components, 84
  and inner product symbol,
    90–91

  phase comparison of, 38, 84–85
  and PRBS signals, 61
  rise and fall time of, 39
  spread spectrum of,
    57–58
  structure of, 38–39
  time domain for, 82, 95
Clock dithering; *see* Spread
  spectrum clocking
Clock nets
  layer stack-ups for, 238
  in power delivery mesh,
    240–241
Code division multiple access
  (CDMA), 13
Collision probability (CP), 305
Conservative field, 150
CP; *see* Collision probability

D

Data signals, 59–60
Device under test (DUT), 37, 137,
  143
Dielectrics
  blurring effect of, 130
  electric field measurements of,
    123, 125
  property, 229
Die power topologies, 231–232,
  239
Differential clock, with and
  without skew, 107–108

Differential symbols
  analysis of, 107
  with skew, 110–112
  time derivative of, 107–108
Digital multitone (DMT) systems,
    307
Digital signal processing (DSP),
    306
Dipole antenna
  charge motion in, 161
  photon emission in, 161
Dipole radiator, near-field and
    far-field transition of,
    133–134
Displacement current, 155
Display frame, 77
  logical structure of, 70
  symbols in, 69, 92
Display symbols
  analysis of, 92
  as complex bit sequences,
    67
  differentiated, 97–99
  energy distribution of, 71
  Fourier components of, 83,
    87–88
  Fourier space of, 86
  for HDMI encoding, 85
  non-differentiated, 100
  radiated emissions from, 104
  spectra of, 66–68, 75, 77, 81,
    97–99
  temporal order impact on, 93
  time domain for, 72, 82, 94
Dithered clocking; see Spread
    spectrum clocking
Divergence-derived surfaces, 190
DMT systems; see Digital
    multitone systems
DSP; see Digital signal processing
DUT; see Device under test
Duty cycle, 46, 297
  harmonics for, 53–54
  repetitive signals, 56
  with rise and fall times,
    297–299

spectrum variation with, 48–52
  trapezoidal signal, 298
DVD player, FCC measurement
    for, 26–27

E

Edge rate asymmetry, 41, 45
Edge rate jitter, 113
  differential signals with, 114
  in symmetric signal, 118
Elastomers, 272, 277
Electric dipole, 165–171
  field impedance of, 172–174
  radial electric field component
    of, 168, 171
  transverse electric field
    component of, 169–171
Electric dipole moment, 167
Electric field distributions, 131
  at even and odd harmonics, 129
Electric field probes, 124
Electric scalar potential, 156–157
Electrolysis, 273
Electromagnetic compatibility
    (EMC), 5, 13, 51, 141, 170
Electromagnetic equations, 149
Electromagnetic interference
    (EMI), 1, 5, 40, 57, 86, 301
  jitter impact on, 113
  limits, 3
  vs. wireless, 2–4
Electrostatic conditions, 151,
    153–154
Electrostatic field, curl of, 152
EMC; see Electromagnetic
    compatibility
EMI; see Electromagnetic
    interference
Extended dipole radiator, 174
  analytical surface for, 180
  field impedance of, 176–179
  radial electric field component
    of, 176
  radiated magnetic field
    of, 175

transverse electric field
    component of, 175

F

Faraday shield; see Shielding
FCC; see Federal Communi-
    cations Commission
Federal Communications
    Commission (FCC), 2–3,
    26–27
Field impedance
  absolute value of, 173
  of electric dipole, 172–174
  of extended dipole, 176–179
  real component of, 172, 178
Flip-chip packages, 230, 241–242
Fourier filtering, 187–188
Fourier transform, 39
Frequency content, 297–301
Frequency planning, 291–297

G

Galvanic compatibility, 273
Gaskets, 273–275
Gaussian noise, 8, 11–12
5 Gb/s interconnect bus, 21
  isolation levels for, 22
GMCH; see Graphic and memory
    control hub
Graphic and memory control hub
    (GMCH), 123
GSM bands, 30
GTEM, 37, 137–139, 207, 300
  of gridded vs. solid power
    planes, 233
  of swapping package power
    layers, 231, 233–234
  test boards, 223, 231, 236

H

Hard disk
  horizontal near-field scan
    of, 125
  vertical near-field scan
    of, 126

Harmonic amplitude, 45–46, 50, 299
Harmonic energy, 40, 48, 69, 140
Harmonic oscillators, 46
Harmonics, 39, 292–293
  and channel skew, 210
  collision, 293, 295–296
  for duty cycle signal, 53–54
  even, 43–45, 77–78, 84, 113–115, 129
  odd, 43–44, 77, 84, 114, 129
  rise time affecting, 46–47
  for symbol, 99–100
  variation with time, 40–43
HDMI connector; *see* High-definition media interface connector
Heat sink, 222–223, 288
Heat spreaders, 222
  circular, 228
  floating, 226, 228
  grounded, 226
  hexagonal, 228
  near-field scan of, 226–227
  for radiated emissions, 225
  shapes for, 224
Hershey's Kiss modulation, 16, 302–303
Hertzian dipole; *see* Electric dipole
High-definition media interface (HDMI) connector, 28, 213
  and data pair skew, 216
  3D model of, 213
  forwarded clock, 216
  and high-speed data pairs, 215
  radiated emissions from, 215
High frequency spatial noise, 186
  smoothing of, 187–188
Horn antenna, 30, 136

I

ICH; *see* I/O control hub
IC types, near-field scanning of, 127

Impulse
  spectrum of, 56
  time behavior of, 57
INFS; *see* Intel near-field scanner
Intel near-field scanner (INFS), 123, 186, 218, 236
Interference mitigation system, 307
Interference source, analysis of, 21–26
I/O control hub (ICH), 123
  vertical and horizontal scans of, 126

J

Jitter, 63, 114–119
  in edge rate, 113
  in pulse width, 113

L

Lagrangian property, 192
Laplacian-derived radiation surfaces, 190
Layout, 281–282
LCD panel
  absorbers in, 279
  antenna placement in, 282
  near-field scans of, 283
Low-level spatial noise, 189

M

Magnetic dipole, 180–185
  far field component of, 182
  near-field component of, 183
Magnetic dipole moment, 180–181
Magnetic field distributions, 131
  at even and odd harmonics, 129
Magnetic field probes, 124
Magnetic vector potential, 156–158
Magnetic wave impedance, 182
  imaginary component of, 182, 184
  real component of, 182, 184
Magnetostatic conditions, 152

Maxwell's equations, 148
  first and second equations, 151, 153
  for static conditions, 150
  third and fourth equation, 153–155
  time-varying conditions, 154–156
Metallic coating, for shielding, 268-269
Microstrip packages, 242–245
Microwave absorbers, 278; *see also* Absorbers
MIMO radios; *see* Multiple input multiple output radios
Minimal mean square error (MMSE), 307
Mitigation techniques; *see* Active mitigation; Passive mitigation
MMSE; *see* Minimal mean square error
Multiple input multiple output (MIMO) radios, 284

N

Narrowband noise, 12
Near-field scanning (NFS), 121, 265
  of hard disk, 125–126
  of heat spreaders, 227–228
  of IC types, 127
  of LCD panel, 283
  of PCBs, 122
  of silicon, 122
  of system clock device, 122–123, 128, 130
NFS; *see* Near-field scanning
Noise
  broadband, 13
  estimation, 306–308
  narrowband, 12
  on production notebook, 4
  quasi-band, 13
  in silicon substrate, 245–251
  wideband CDMA, 13
  thermal, 17, 19

Non-Gaussian noise, 7, 11, 307
Notebook
  narrowband noise in, 12
  noise measurements on, 4, 11
  wideband CDMA noise in, 13

O

OATS; *see* Open area test site
OFDM; *see* Orthogonal frequency
    division multiplexing
Ohmic losses, 256
Open area test site (OATS), 141
Orthogonal frequency division
    multiplexing (OFDM), 304,
    306

P

Packages, 121; *see also* Flip-chip
    packages; Wire-bond
    packages
  electric field energy distribution
    for, 243
  microstrip, 242–245
  power distribution radiated
    emissions in, 230
  RFI potential of, 229, 231
  strip-line, 242–243, 245
Packet error rate (PER), 307
Passive mitigation
  absorbers, 276–281
  layout, 281–282
  shielding, 254
PCB; *see* Printed circuit boards
PCIe, 19, 196–197, 217
  data stream measured on, 60,
    62–64
PCIe Mini Card, 24–25
PDN; *see* Power delivery network
Penetration loss; *see* Absorption
    loss
PER; *see* Packet error rate
Phase locked loop (PLL), 186, 190
Photons, 151, 158–159
Platform components, spectra of,
    26

Platform noise, 6–8; *see also*
    Noise
  ADC with and without, 9
  characteristics of, 11–14
  in GSM bands, 30
  from interference system, 9–10
  on PCB, 32–33
  on production notebook, 11
  in radio bands, 17, 20, 69, 292
  risks, 17
  wideband measurements of, 26
  in WiMax band, 31
  on wireless systems, 9–10
Platform signals, spectral analysis
    of, 38–40
PLL; *see* Phase locked loop
Point sources, 189, 196–197, 199
Power delivery network (PDN),
    130, 186, 188
  emissions testing of, 231
Power distribution radiated
    emissions, 215
  in packages, 229
  in silicon, 229
Power plane
  gridded *vs.* solid, 232–233
  swapping of, 234
Power plane resonances, 217
  isolation comparison due to,
    221
  radiated power distributions
    of, 220
PRBS signal; *see* Pseudorandom
    bit stream signal
Printed circuit boards (PCB), 29
  form factor of, 218
  measurement of, 221
  near-field scanning of, 122
  platform broadband emissions
    from, 32, 137
  platform noise on, 32–33, 138
  power delivery in, 215
  resonance structures in, 217
Pseudo-BGA, 223–224, 231, 237
Pseudorandom bit stream (PRBS)
    signal, 31, 59–60

clock and, 61
differential model of, 196,
    201–203
spectra of, 61, 65–66, 77, 81
time variation of, 74
Pulse width jitter, 113
  in symmetric signal, 118

Q

Quasi-band noise, 13

R

Radial electric field component
  of electric dipole, 167–168
  of extended dipole, 176
Radiated emissions
  from display symbols, 104
  from HDMI connector, 215
  for heat spreaders, 225
  between single and multiple
    pairs of sources, 226
Radiation mechanisms, 158–162
  in conductor, 162
  of silicon, 188
  from transmission line, 162
Radiation sources, 123, 199–200
Radiation surfaces, mask value
    and, 190–191, 194–195
Radio bands
  harmonic collision in, 293
  interference levels in, 18
  platform noise in, 17, 69, 292
  spread clock in, 303
Radio frequency interference
    (RFI), 5–6
  of clocks, 114
  frequency planning in, 254
  jitter impact on, 114–119
  of package, 229, 231
  risks, 17
  of symbols, 114
Radio platform channel model, 8
Reflection attenuation, 278
Reflective loss, 256
  of aluminum, 261
  of copper, 261, 265

far-field, 264
near-field, 263–264
Reverberation chamber, 142–145
RFI; *see* Radio frequency
interference

S

SAC; *see* Semi-anechoic chamber
SE; *see* Shielding effectiveness
Semi-anechoic chamber (SAC),
far-field measurements of,
141–142
Shield discontinuities, 269
Shielded cavity, 276
Shield impedance, 261
Shield thickness, 268
Shielding, 254
characteristics, 257, 260
design, 255
far-field case of, 263
metallic coating for, 268–269
multi-reflection effect of,
266–267
near-field case of, 263
Shielding effectiveness (SE), 255,
270
of copper, 262
of seams, 272–273
vs. frequency for aperture, 271
Signal-to-interference (SIR), 8
Signal-to-noise ratio (SNR), 11,
20, 274, 306
Silicon, 37
guard ring structures in, 249
near-field scanning of, 122
power delivery in, 237
power distribution radiated
emissions in, 229
radiation mechanisms of, 188
substrates, noise coupling in,
245–247

Single-ended signals, and
differential signals, 21–22,
111
SIR; *see* Signal-to-interference
Skin depth, 256–257
of aluminum, 257–258
Slot antenna, 270
SNR; *see* Signal-to-noise ratio
Source interference, multiple
channels of, 205
Spread spectrum clocking (SSC),
13, 301–305
frequency domain plot of,
301–302
Hershey's Kiss modulation
profile, 16
linear and non-linear
modulation, 302–303
modulation frequency of, 16
and non-SSC clocks, spectral
frequency of, 14–15
in radio band, 303
in reducing peak amplitudes,
29
time domain response of, 301
triangular modulation profile,
16
SSC; *see* Spread spectrum
clocking
Step function
spectrum of, 58
time behavior of, 59
Stokes theorem, 153
Strip-line packages, 242–243, 245
Susceptibility testing, 142
Symmetric clock, 85
odd harmonics, 44
time and harmonic number for,
41–42
and asymmetric clock, phase
shift of, 84–85
Symmetric trapezoidal signal,
139–140

System clock device
analytical structures for, 185,
189–190
electric field measurements of,
122
horizontal scan of, 128
near-field scanning of,
122–124, 128, 130, 185
radiated emissions from, 139,
141
vertical scan of, 129
without package dielectric
effects, 131
SystemView, 196, 203, 208

T

Thermal noise, 17, 19
Time-varying Maxwell's
equations, 154–156
Tone puncturing, 308
Transverse electric field
component
of electric dipole, 169
of extended dipole, 175

V

VLSI GTEM, 37, 137–138

W

White Gaussian noise, 7
Wideband CDMA noise, 13
WiMax, 29, 31
Wire-bond packages, 229–230,
241–242
Wireless channel, interference
potentials in, 104
Wireless receiver bandwidths, 19
WLAN, 104, 262, 294
antenna placement, 283, 286
time domain in, 59
WWAN, antenna placement in,
289–290

Printed and bound by CPI Group (UK) Ltd, Croydon, CR0 4YY

03/10/2024

01040333-0003